Foreman v. Xymos

An Artificial Intelligence and Nanotechnology Trial
based on the science fiction novel, *Prey,* by Michael Crichton

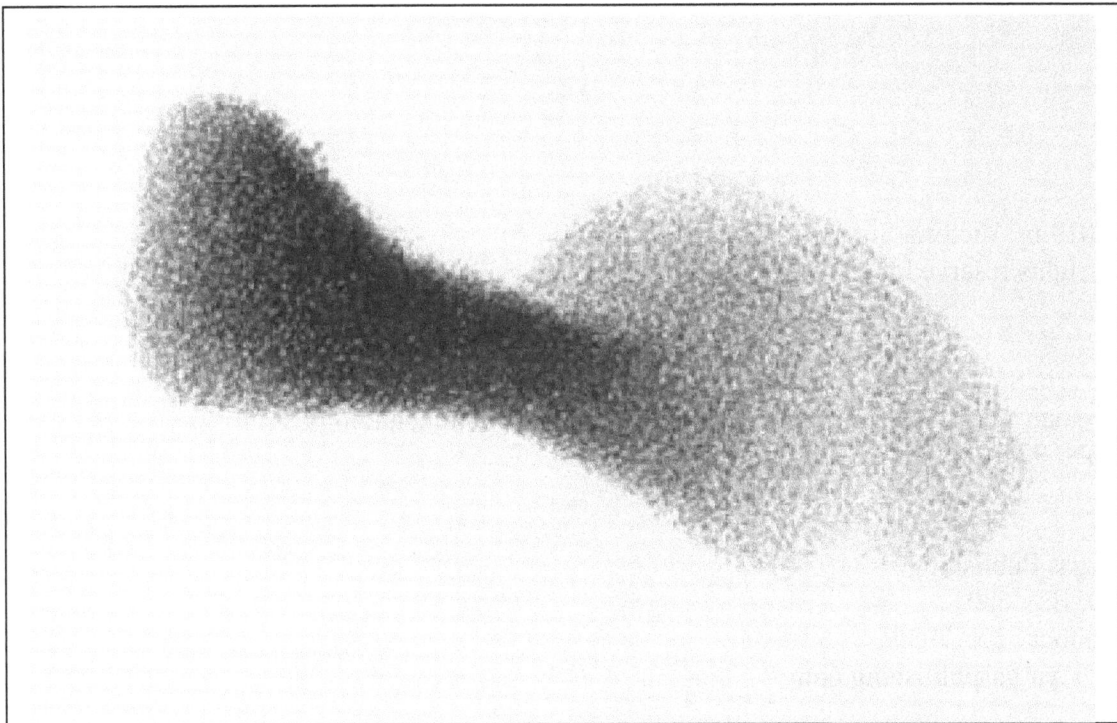

Victoria Sutton, MPA, PhD, JD
Paul Whitfield Horn Professor
Texas Tech University School of Law

Victoria Sutton
Foreman v. Xymos, An Artificial Intelligence and Nanotechnology Trial
ISBN: 978-0-9968186-5-0

Vargas Publishing
P.O. Box 6801
Lubbock, TX 79366
www.vargaspublishing.com

To Remington and Summer

TABLE OF CONTENTS

ACKNOWLEDGMENTS

This trial practice exercise is an application of knowledge and skills from my Nanotechnology Law and Policy course, taught at Texas Tech University School of Law since 2009. I thank my students who took my course and performed this trial practice exercise with enthusiasm. My students have also contributed to the shaping of my casebook, Nanotechnology Law & Policy (2010) Carolina Academic Press. Pioneering this uncharted area of law practice addressing nanotechnologies and its impact on our society is daunting but exciting, and I believe, prepares law students for the world they will be facing in the very near future.

I also owe much gratitude to Professor Anthony Morella who was the first to demonstrate to me that trial practice could be taught --- and learned, when I was enrolled in his trial practice class at American University, Washington College of Law.

Also, I wish to acknowledge the following for the use of their publications:

Crump, David, "Basic Methods of Qualifying or Introducing Some Common Items of Evidence," Texas Bar Journal 702-3 (July 1980).

Snow, Jennifer and Giordano, James, "Commentary, Aerosolized Nanobots: Parsing Fact from Fiction for Health Security—A Dialectical View," *Health Security,* 17:1, 2019, Mary Ann Liebert, Inc. DOI: 10.1089/hs.2018.0087. [Excerpt].

Moskovits, Martin. "Nanoassemblers: A Likely Threat?" 4 Nanotechnology L. & Bus. 189, Nanotechnology Law & Business, Legislative & Regulatory Affairs (June 2007). [Excerpt].

I. INTRODUCTION

This is a story of the experimental use of artificial intelligence and nanotechnology to develop a weapon for the U.S. military. It is based on the narrative in the science fiction novel, *Prey*, by Michael Crichton. In fact, Crichton sets up the call to justice at the end of the book, when his protagonist talks about the lawsuits and criminal cases that will be coming.

Michael Crichton has always been a signal human for society, noting with literary flare when a technology holds potential horrific consequences, and then builds his idea through a thriller-style novel to show us exactly how it might happen. This is exactly what he did in *Andromeda Strain* (1969) which told the story of a disaster from contamination from outer space. Then again, in *Jurrassic Park* (1990), he told the story of what would happen if everything we dreamed possible with genetic engineering might actually come true, and the horrific consequences of doing it. Crichton takes up the subject of artificial intelligence and nanotechnologies in *Prey*, and develops a story that shows what would happen if we actualize our vision for these technologies, without the constraints of law, morals or ethics.

Andromeda Strain was made into a movie in 1971, and it was one of the top ten grossing movies for the year, and continues to be a classic. *Jurassic Park* was a best seller as a novel, and resulted in several sequels in the 1990s to 2001; and a remake of the story as *Jurassic World* in 2015. Once his books get past the threshold into the theater, it seems they are unstoppable. Yet, *Prey* (2002), another promising technological disaster thriller, never made it to the theater. While there were talks and apparently some contractual commitments, the movie production fell apart, and ended the possibility of a theatrical version of *Prey,* at least for now. Sadly, Crichton passed away in 2008.

This book includes a trial casefile based on the facts in the book, and develops the needed expert witnesses, building on the fact witnesses in the narrative. The disaster capitalizes on the fears of military weapons development in secret in government contractor laboratories, grey goo, and the risk of unleashing the unknown into the environment. I would like to think Crichton would have enjoyed seeing how the harms he portrayed in the book were analyzed by the court, in response to a technological disaster involving artificial intelligence and nanotechnologies.

II. APPROACH TO TRIAL PRACTICE

Becoming familiar with the materials in the entire case file is the first step. Read through the entire case file and think about how you would develop a litigation strategy or defend against a law suit in this scenario. You will have assignments related to various parts of the trial, for example, pretrial motion drafting, arguing pretrial motions, a direct examination, a cross examination, an opening or closing statement.

If you are preparing expert witnesses, you will need to spend some time with them in preparation for their direct examination and anticipate what questions you will need on cross examination. The opposing party, of course, should not talk to the expert witness and will have only the benefit of the expert's deposition to design a cross examination of the opposing expert. Each party will have their own experts. An experiences expert witness knows that he/she does not have to talk to the opposing party (and should not talk to them) except in a sworn deposition. So for purposes of any exercise on this phase of the trial, the opposing party should not talk to the opposing experts outside of the courtroom.

If this is being done live, with a class in a courtroom, students who are not participating in the immediate courtroom activity, should be seated in the jury box in order to provide a realistic jury for students to address.

The time limits for the trial will be as follows:

Opening statements: ____minutes

Direct examination: ____minutes

Cross examination: ____minutes

Closing statements: ____minutes

The order of the trial will be as follows:

Pretrial:
Motions for Summary Judgement and Motions in Limine----Defendant
Motions for Summary Judgement and Motions in Limine --- Plaintiff

For a Criminal Trial
Motions in opposition to indictment --- Defendant
Motions in support of indictment ---- Plaintiff

Trial:
Opening statement ----plaintiff
Opening statement-----defendant

3

<u>Expert Witnesses</u>

Plaintiff's expert witness:

    Dr. Roden.

        Voir dire -- plaintiff's attorney

        Direct examination – plaintiff's attorney

        Cross examination ---- defendant's attorney

        Re-direct examination --- plaintiff's attorney (optional)

        Re-cross examination --- defendant's attorney (optional)

        Motion for directed verdict – defendant's attorney

Defendant's expert witnesses:

    Dr. Lindman

        Voir dire --- defendant's attorney

        Direct examination --- defendant's attorney

        Cross examination ---- plaintiff's attorney

        Re-direct examination --- defendant's attorney (optional)

        Re-cross examination --- plaintiff's attorney (optional)

        Motion for directed verdict --- plaintiff's attorney

Closing argument ---- plaintiff

Closing argument ---- defendant

III.   WITNESS PREPARATION

It is essential that the teams spend some outside class time with their expert witnesses to prepare them for not only direct examination, but for anticipated questions on cross examination. In order to prepare the expert witness, it is essential that the expert become familiar with their own resume, and the types of evidence that they will be expected to explain. It becomes painfully obvious when a student has failed to prepare their expert witness for the trial practice, and may be a hard-learned lesson.

The attorney should have prepared questions for direct examination and should go over those questions with the expert witness. A good trial attorney should never ask a question that he/she does not know what will be the answer, or so they say. You will have to make that call. Only with substantial experience should an attorney venture into the unknown with a line of questioning. The purpose of this rehearsal is not to tell the expert what to say, but rather to make sure that the question elicits the type of evidence that the trial attorney needs to establish every element of the case. Failure to make the relevant expert testimony a part of the record may result in a grant of a motion for a directed verdict at the end of the evidence.

In preparation for cross-examination, the attorney should review the anticipated questions that will arise for the expert witness based upon problem statements in the deposition or problems with flaws in the scientific evidence and exhibits. This, too, is important because after the cross-examination, there will be no time to explain rehabilitation or to prepare the expert about what must be explained for the record, when the attorney has only one chance for re-direct examination.

Whether this expert is a law student or a science student, the experience will be beneficial to take the role of the expert both in preparing and in testifying.

IV.   PRE-TRIAL MOTIONS

The pre-trial motions include the motion *in limine* and the motion for summary judgment.

The motion *in limine*, includes the objection to the testimony being offered or objection to the qualifications of the expert for the type of testimony that the expert expects to proffer. The motion can address any or all of the experts and their testimony.

The motion for summary judgement requires that the court dismiss the case with prejudice against the party, which has failed to carry their burden of showing that evidence will be introduced to prove every element of their case. Failure to have any expert survive a motion *in limine* may result in the elimination of the testimony which would support one of the essential elements of one or more of the causes of action in this case.

V.   VOIR DIRE FOR EXPERTS

The voir dire is conducted by the plaintiff of the plaintiff's expert witnesses and by the defendant of the defendant's expert witnesses. This process typically begins with the attorney asking the expert witness a series of questions about their background, for example,

"What was your undergraduate major?"

"In what subject do you hold a graduate degree?"

"Is that a Ph.D. degree?"

This is all to lay a foundation for the qualifying your expert witness so that they may testify as an expert and render opinions on particular issues for which they have expert training or knowledge.

This process serves another purpose, as well as putting qualifications on the record. The voir dire provides the first opportunity for the expert witness to meet the judge and jury and this is the time to make first impressions on both. The attorney should draw out answers which will allow the expert to make eye-contact with the jury, perhaps; and to allow the jury to gain trust and confidence in the expert witness. The voir dire also serves the purpose of providing a warm-up for the attorney to practice an interchange with the expert witness before the critical questions are necessarily presented for the expert witness's response.

A resume for the attorneys is included in this casefile, so your expert will want to become familiar with their resume for this phase of the trial. You must share the resume of the expert with the opposing party before trial. You will also need to provide it to the opposing party which you enter the expert's resume into evidence as an exhibit which you will want to do.

VI.   DIRECT, CROSS, REDIRECT AND RECROSS EXAMINATION

Direct examination

Direct examination of an expert witness requires the attorney to elicit the required testimony and opinions from the expert witness for the record. Prior preparation is essential and the practice of reviewing the questions with the expert will help ensure that the trial strategy will be effective. Questioning may not be leading, and the testimony must come from the expert witness, not from the attorney.

Cross examination

Cross examination is the opportunity for the opposing attorney to examine the expert witness on weaknesses in the testimony in an effort to discredit the expert's credibility or to discredit the expert conclusions. Questions may be leading, and should be leading, eliciting a "yes" or "no" response, only, when possible. Keeping control of the questioning is an important strategy on cross-examination.

Redirect examination

Redirect examination is not required, but if the expert witness has been impeached, this allows an opportunity for the attorney to rehabilitate the witness. Here, the questions must not be leading, and may be risky if the expert witness is not prepared for this contingency.

Recross examination

Recross examination is rarely done, and may not be permitted. Recross may be allowed for one or two questions if necessary to clarify the testimony of the expert witness.

VII.   IMPEACHMENT AND REHABILITATION

Impeachment of any expert witness will require the opposing attorney to rehabilitate the expert witness, if the attorney decides this is possible at the close of cross-examination. It is during re-direct examination that the attorney has the opportunity to repair any damage done to the testimony or credibility of the expert witness during the cross-examination.

In impeachment by prior inconsistent statement, the opposing attorney may confront the witness during cross examination if the issue was raised in the direct examination. If the witness denies making the inconsistent statement, then the document or evidence may be introduced for purposes of impeachment. Federal Rule of Evidence 613(b) provides that the witness must be "afforded an opportunity to explain or deny" the prior inconsistent statement in order to allow the admission of evidence that will be used to impeach this witness.

Any witness may be damaged during cross-examination and require clarification of testimony, which is a form of rehabilitation.

If rehabilitation fails, then the impeachment of the testimony of that expert witness may prove devastating to the case.

VIII. INTRODUCTION OF EXHIBITS

The introduction of exhibits is another skill which is important to any trial. The introduction of exhibits requires that the attorney lay a foundation for the introduction by identifying the evidence and why the expert witness knows what it is. This establishes that the document is what it purports to be, for example, the reports and depositions of the expert witnesses.

The attorney introducing the exhibits must remember that in laying the foundation requires that the testimony be both authentic and relevant and comply with hearsay rules and the best evidence rules. Authenticity is simply that the document or evidence is what it purports to be. Relevance means that the evidence must link with an issue you want to prove. It does not have to prove the entire issue, but may be only a brick in the wall of proving an issue.

The document must be submitted to the court (a clerk or the judge) and marked and numbered as an exhibit. The attorney may number the documents before the hearing, but this does not allow for documents that may be excluded, or changes in strategy or order. Thereafter, that testimony should be referred to by exhibit number, so each attorney should keep a list of the exhibits and the numbers assigned by the court.

The attorney should then ask the witness to identify the document. The attorney may hand the document to the witness for examination. When the witness has responded, the attorney may then ask something about the nature of the evidence --- is this document the deposition you gave on March 3, 2019, for example.

Introduction to Introducing Exhibits

1. Introducing a Photograph into Evidence

Step 1:   Have the exhibit marked by the court reporter.
Step 2:   Have the witness identify it and lay the predicate for its admissibility. (NOTE: it is unethical to expose inflammatory exhibits to jury view before they are admitted. Some trial lawyers do so routinely, for a simple reason: they are unethical)
Q: Mr. Plaintiff, I show you what has been marked plaintiff's exhibit no. 1, and I ask you whether you can identify it. Can you?
A: Yes, I can.
Q: What is it?
A: It's a photograph showing the damage to my care after the accident I've just told about.
Q: And does it truly and accurately reflect what the car looked like at that time?
A: Yes, it does.

Step 3:   You have now laid the predicate for introduction of the exhibit. You should now walk over to opposing counsel and tender it to him/her.

Step 4:   Formally offer the exhibit into evidence. "Your honor, at this time I offer plaintiff's exhibit no. 1 into evidence.

Step 5:   Your opponent may now object. You should be ready to make a short argument in support of your exhibit. "Your Honor, even though the cost of repairs have been stipulated as my opponent says, the exhibit is still relevant to the way in which the accident happened." You should make this argument only if it seems appropriate, that is, you should remain quiet if the judge overrules the objection away.

Step 6:   If an objection is sustained and you think the evidence is both admissible and important, ask further questions directed to laying the predicate for it.

Step 7:   Be sure the exhibit is formally received by the judge.

Step 8:   If it seems desireable, ask to have the exhibit passed by the bailiff to the jury or ask the witness to read excerpts from it if it is a document.

2.   Introducing a Tangible Object
       (Note: the steps outlined above for marking the exhibit, offering it, etc. apply generally to objects and documents. They will not be repeated here but should be done. The predicate is as follows:)
       Q: I show you what has been marked as plaintiff's exhibit 2 and ask you whether you can identify it. Can you?
       A: Yes.
       Q: What is it?
       A: It is the actual steering wheel of the car I was driving at the time.
       Q: How do you recognize it?
       A: [Here there can be a variety of answers. "Because I drove that car for 3 years and I know its appearance." "Because it contains a crack at the top where I hit my head in this accident." "Because my initials and the date of the accident are scratched on the side of it, here, and I put them there when it was removed from the car."

A word to the wise: Tangible objects whether they be guns, bent fenders, leg casts, fingerprints, etc., often have an evidentiary value greater than their relevance to the issues in the case; and even if you have had the cracked steering wheel described thoroughly by testimony it will probably enhance your case to introduce it.

3.   Introducing a Business Record
       Business records are hearsay. They constitute out-of-court statements offered to prove the matter stated in them. However, there is an exception to the hearsay rule for business records meeting certain qualifications. The questions and answers that would qualify them might be as follows (after the witness' name and background has been given):
       Q: Do you have actual care, custody and control of the business records of Joe's Auto Shop?
       A: Yes. [A qualified witness other than the custodian may sometimes be proper, but to be sure, get the custodian.]
       Q: Are these records kept and prepared by Joe's Auto Repair in the ordinary course of business?

A: Yes.

Q: Are the entries made at or about the time of the events they record?

A: Yes.

Q: Are they made initiatlly by someone who has personal knowledge of the events recorded?

A: Yes. [Again, personal knowledge of the entrant is not always required, but a "yes" answer here makes introduction easier.]

Note: It is a good idea to go over these questions carefully with the witness to ensure that he/she understands them. An auto repairman may get thrown by a mumbo-jumbo phrase such as "at or about the time of the events they record," even though it is required by the business records stattue, and may answer erroneously with the result that introduction fo the records becomes difficult.

Second Note: Once the records are received in evidence you can have the witness "interpret" them. For example, if one of the issues in the case concerned the plaintiff's brakes:

Q: What work , if any, do these records show was done approximately 8 weeks before this occurrence on March 21, 1977?

A: A complete brake job. Here where it says "linings," that's brake linings, and this line here lists the other parts. And there's labor.

Q: Does the record reflect whether the automobile was road tested?

A: It was.

4.   Qualification of an Expert Witness
The expert may be anything from a garage mechanic to a metallurgist. He may be qualified by experience along, by training along, or by both. Question sand answers to do so might go something like this (after the witness' name and background have been elicited:)

Q: What is your occupation or profession?

A: I'm a metallurgist. That means I study and work with metals. [Or: I'm a mechanic at Joe's Auto Shop.]

Q: What qualifications do you have for the work --- what education, training or experience?

A: I have an AB in physics from Harvard and a PhD in metallurgy from the California Institute of Technology. I am on the faculty of the University of Texas, department of physics, where I teach metallurgical courses; I am a member of the American Society of Metallurgists. I worked 10 years in industry and am the author of 25 articles in learned journals.. [Or: I done worked as a mechanic for nigh onto 30 year. I was borned into this line of work and I done fixed thousands of folks' cars. I know just about all there is to know about how to fix brake systems on cars like this one here. It's how I make my livin'.]

Introduce resume of the expert as an exhibit, and then ask the expert to identify it.

Q: Is this a copy of your resume?

A: Yes, it is.

Q: I ask the court to accept Dr._____as an expert in the field of metallurgy.

Note: At this point, the opposing counsel has an opportunity to object, but this is rarely done. Only if the field for which the expert is qualifying is not appropriate or is so unusual that the opposing counsel may want to object.

Once the expert has been qualified, he may be asked (1) a hypothetical question base don facts in evidence in the case, (2) a question based on his own examination of objects involved, or (3) a combination of both.

Q: All right, now, Mr. Mechanic, you say you saw Plaintiff's car after the brakes were fixed. In your opinion, were they fixed properly, so that they'd stop the care normally?
A: Yes, they were.
Q: Then, let me ask you the following question, based on the facts and testimony in this case. If plaintiff were driving 30 mph on a dry asphalt road and applied the brakes suddenly, would the car continue to respond to steering as it stopped?
A: You could still steer it, just like if you wasn't applying no brakes at all.
Q: Then I take it that the car, in you opinion, would not veer sharply into the other land (as one witness has said) if it was steered straight ahead?
A: No. It wouldn't.

Adapted from: David Crump, *Basic Methods of Qualifying or Introducing Some Common Items of Evidence*, Texas Bar Journal, July 1980.

IX. PLOT SUMMARY

Crichton, *Prey*

As you learned in a previous lecture, the novel *Prey*, is based on a scientists' and engineers' opinions about what might happen with the risks of nanobot self-assemblers and the lack of control once their artificial intelligence is unleashed, where their objective is survival.

There were disputes about whether this could ever happen, when the book was published in 2002. But in January 2017, the Department of Defense sent out a news release which announced micro-drones had been successfully demonstrated using swarm behavior artificial intelligence. These are micro-drones, not nano-drones, or nanobots – at least not yet.

The Fact Scenario

- Xymos is a private company, developing swarming nano robots (nanobots, for short) for the military.

- The purpose of these swarming nanobots are to use their cameras for reconnaissance and spying

- The nanobots were created from *E.coli* bacteria which were designed to create gamma assemblers from raw materials in the environment. They were expected to scatter in the wind and eventually decay, when released into the environment.

- Instead, they adapted for survival with their artificial intelligence and swarm behavior, and that became their priority.

********Spoiler alert*******

- The developer of the nanobots, Julia, is led to release them into the environment to continue her experimentation, but without permission.

- This swarm survival behavior drove the nanobots once in the environment to utilize carbon sources to propagate – self assemble from materials in the environment.

- The energy source was solar energy, at least in the beginning.

Plot summary

<u>Act One</u>

- The story is told through the perspective of Jack Foreman, who was once a contractor at Xymos as a computer software engineer, working on artificial intelligence algorithms. His whistleblowing against the company for whom he works, Mediatronics, got him fired.

- His wife, Julia Foreman, is Vice President of Xymos, and is wildly enthusiastic about perfecting a new medical device based on nanotechnology, or so she says.

- They both live in California, while the Xymos lab, the, "fab plant", is in the desert of Nevada near Las Vegas, and they fly to work.

- Jack and Julia live together and are married. Jack, now unemployed, stays at home with their baby, Amanda; Nicole, a preteen daughter; and Eric, the oldest child.

- Strange events begin to occur in the Foreman household. First the baby gets an unusual rash that is worsening, but during the MRI to diagnosis her, the rash instantly disappears and she is cured. No explanation.

- The baby's mother, Julia, sends a Xymos cleanup team to her house and they sweep the baby's room.

- Jack also notices his son's MP3 player memory chip is a pile of dust.

- To make matters worse, Julia is behaving strangely, coming home very late from work, and one night leaves to go back to work shortly after arriving at home. From Jack's view, he sees what looks like the figure of a man in Julia's car.

- When Medtronics calls Jack and offers him his job back, because they have become a contractor for Xymos, Jack takes it. They are hired to solve a computer programming problem for Xymos.

- Jack rejoins his former coworkers from Medtronics at the "fab plant". One former worker, Ricky, is now working for Xymos. Ricky gives Jack a tour of the plant and tells him what their real work is about – they are under contract with the Department of Defense, to create nanobots with artificial intelligence that can perform reconnaissance and spying missions. The swarm is based on engineered *E.coli* bacteria and this is its method of self-assembly to recreate itself.

- Ricky tells Jack that it is the algorithm that he created for artificial intelligence that is causing the problem with the nanobots. It seems the nanobots were accidentally released into the wild through a faulty vent system and they are using solar energy and reproducing from elements in the wild, quickly learning to innovate. They exhibit predatory behavior, attacking and killing animals in the desert. The winds at night keep them away from the plant, but during the day they swarm anywhere.

Act 2

- A dead rabbit is discovered outside, and Mae, one of Jack's former Medtronic coworkers, now also at fab plant, goes outside to inspect with Jack.

- They determine the rabbit died of suffocation from the nanobots filling its brochial tubes. Jack is attacked by the nanobots when Mae is briefly inside to get some equipment. Jack struggles to get back inside the airlocked facility, but suffers anaphylactic shock from the attack, but survives.

- The team needed persuading by Jack to destroy the swarms, and when they try to destroy them, they get another surprise – the nanobots are intent on surviving.

- While outside the airlock in a shed, two of Jack's coworkers, David, an engineer and Rosie, a natural language processing specialist, are both attacked and killed by the nanobots.

- The rest of the team, Bobby, a programming supervisor; Charley, a programming specialist in genetic algorithms; Jack and Mae take cover in the cars that are nearby, but the swarms soon enters the cars.

- Jack and Mae escape to the lab because fortunately, the wind started blowing, causing the nanobots to blow elsewhere. Charley tried to spray them before running but was overcome by the attack. Jack ran back and saved Charley when no one else would go.

- From inside the lab, they are able to determine the swarms are innovating and evolving. The nanobots are tracking and tracing the movements of their prey, photographing the prey with their cameras and swarming to form rough 2-D images of the prey for the swarm.

- The swarms are also capable of forming a 3-D image of their prey, revealed to the group when they saw an image of Ricky, formed by the nanobots in the desert.

- The bodies of the next two fatalities, Rosie and David were carried away by the nanobots at night.

- Jack, Mae and Bobby head out into the desert to track the swarm and catch up with the swarm moving Rosie's body and follow it to a cave where the swarms are nesting. The swarms come out to attack Jack, Mae and Bobby.

- Again, serendipitously, a Xymos helicopter arrives. The thrashing helicopter blades blow the attacking nanobots back into the cave.

- The helicopter-generated wind keeps the nanobots at bay, while Jack, Mae and Bobby go into the cave to plant some thermite to blow it up. They observe the nanobots have an assembly plant that looks a lot like the plant where they were "born".

- Julia, having been released from the hospital, greets them back at the fab plant.

- Charley is found dead with a nanobot swarm around him, and the communications lines have been cut, and Julia indicates that Charley was trying to cut the lines.

- The next morning, Mae and Jack see on the surveillance video that they are the only ones left, who are not infected. Julia and Ricky were having an affair. Charley was in a fight with Vince, the maintenance guy; and Ricky in the communications room. They see Julia kiss Charley, swarming the nanobots into his mouth.

- Jack concludes that he can stop the nanobots by destroying the *E.coli* assembler with a phage that eats bacteria. His dissemination device will be the sprinkler system in the lab.

- Jack and Mae drink the phage solution and Mae has diverted the phage solution into the sprinkler system.

- But he is met by fierce opposition by Julia, Vince and Ricky who are "possessed" by the nanobots and are in survival mode. To stop the sprinkler, Ricky turns off the plant safety system which would trigger the sprinklers. But will cause the sprinkler system to overheat, and it explodes, so Julia and Ricky reactivate the sprinkler system and know they will be drenched, but they are dead either way.

- Jack and Mae escape before the plant explodes due to a methane leak and thermite Mae had in the building.

- Jack takes the phage solution he kept home to give to his children.

- Mae reports the incident to the U.S. Army.

- Jack realizes his wife had been infected with the nanobots after training them with toys and entertaining them, she had become affectionately attached to them, and they took over her body and mind.

- Jack goes through her email and finds the release of the swarm into the environment was done intentionally by Julia. She thought giving them a chance to evolve would work out the AI problem, but she had not considered the consequences of the release.

Denouement (the wrap up)
Here, Crichton sets up the narrative for our trial, with his protagonist, Foreman, reflecting on what might happen next with regard to criminal and civil liability for Xymos and Larry Handler.

- Foreman: "The Army is acting dumb about this whole thing, but I have Julia's computer here at home, and I have an email trail on her hard drive. I removed the hard drive, just to be safe. I duped it, and put the original ina safe deposit box in town. I'm not really worried about the Army. I'm worried about Larry Handler and the others at Xymos. They know they have horrific lawsuits on their hands. The company will declare bankruptcy sometime this week, but theyr're still liable for criminal charges. Larry especially. I wouldn't cry if he went to jail."

Major Characters

- Jack Forman - A former team lead/manager at MediaTronics, who later works at Xymos as a contractor

- Mae Chang - A field biologist on Jack's consulting team.

- Ricky Morse - A friend of Jack's, works for Xymos

- Charley Davenport - A member of Jack's team who specializes in genetic algorithms.

- David Brooks - engineer on the team.

- Rosie Castro - specialist in natural language processing.

- Bobby Lembeck - A programming supervisor.

- The "Swarm" is any of a number of funnel-clouds of predatory nanobots that consume living things. It can learn and adapt and transmit to the next generation what it has learned through its *E.coli,* replicating mechanism.

- Vince Reynolds - maintenance director, the Xymos lab.

- Amanda- Jack and Julia's infant daughter.

- Nicole - Jack and Julia's preteen daughter.

- Eric - Jack and Julia's son.

- Julia - Jack's wife, Vice president of the Xymos company, whose body and mind is invaded by the nanobots.

Minor Characters

- Ellen - Jack's sister from out of town. She is a caregiver for Jack and Julia's children while he is in Nevada. She is suspicious of Julia's behavior and believes she is taking some kind of stimulant.
- Don Gross - Jack's former boss, who fired Jack.
- Gary - Jack's lawyer.
- Maria - Jack and Julia's cleaning lady.
- Annie - Jack's headhunter.
- Carol - Julia's Assistant.
- Mary - Ricky's wife.
- Tim Berman - The man that took over Jack's job.

IN THE UNITED STATES DISTRICT
COURT FOR THE DISTRICT OF
NEVADA

JACK FORMAN, MAE CHANG, ESTATES ⟩
OF CHARLEY DAVENPORT, VINCE ⟩
REYNOLDS, JULIA FORMAN, ⟩
NATURALIST IN PARK, AND ⟩
PETER MORRIS ⟩
     Plaintiffs, ⟩
⟩
     VS. ⟩
⟩
XYMOS, INC., LARRY HANDLER, CEO ⟩
AND U.S. DEPARTMENT OF DEFENSE ⟩
     Defendants. ⟩

Cause No. 01-123123

Plaintiff's Original Petition

1.   Plaintiffs are Jack Forman, Mae Chang, Estates of Charley Davenport, Vince Reynolds, Julia Forman, Naturalist in the park, and Peter Morris.

2.   Defendants are Xymos, Inc., Larry Handler, CEO, and U.S. Department of Defense. Xymos

3.   Venue is proper in The United State District Court for the District of Nevada in the State of Nevada that is the site of the occurrence of the injuries suffered by plaintiffs.

4.   Plaintiffs suffered severe trauma, death, damages, and emotional distress as the real and direct result of the release of the nano particles ("the swarm") into the environment.

5.   The defendants have a duty to act with reasonable care and adhere to the laws/regulations of Nevada, Environmental Protection Agency ("EPA"), Comprehensive Environmental Response, Compensation, and Liability Act ("CERCLA"), and other applicable laws and regulations.

6.   Defendant Larry Handler is an employee and authorized agent of Xymos, Inc.

7.   Defendant U.S. Department of Defense is subject to suit under the Federal Tort Claims Act.

Jurisdiction

The district court has jurisdiction of the case that is docketed as No. 01-123123 pursuant to 28 U.S.C. § 1331.

25

Causes of Action
Negligence/Negligence Per Se

8. Defendants owed a duty to Plaintiffs which they breached and which proximately caused the damages complained of herein.

Toxic Tort

9. Defendants knew about, were responsible for, and did not take measures to prevent the release of the swarm into the environment.

10. Defendants intentionally released the swarm into the environment which was the proximate cause of the damages complained herein

Nuisance

11. Defendants directly, and through the intentional and reckless/negligent release of the swarm, unreasonably and substantially invaded the property of the Plaintiffs interfering with the comfortable use and enjoyment of said property.

Wrongful Death

12. Defendants, through the intentional and reckless/negligent release of the swarm, were the proximate cause of the deaths of Charley Davenport, Vince Reynolds, Julia Forman, Ricky Morse, Rosie Castro, Bobby Lembeck, and David Brooks. Plaintiffs are survived by spouses, children, and/or other dependants who were substantially and irreparably harmed as a result of said intentional and reckless/negligent acts.

Strict Liability

13. Defendants engaged in the design, manufacture, and intentional release into public of the swarm that is the subject of this lawsuit.

14. At all times mentioned herein, and at the time of the design, manufacture, and release of the swarm, the swarm was defective and not reasonably safe and fit for the purpose intended nor for reasonable intended purposes, uses or operations; was defectively designed; was defectively manufactured; failed to provide adequate and proper warnings of the hazards and dangers associated thereto; one or more which was the proximate cause of the aforesaid injuries to Plaintiffs.

Trespass to Land

15. Defendants directly and through its intentional and reckless/negligent release of the swarm entered the land of Plaintiffs without authorization causing damage to said property.

26

Assault/Battery

16. Defendants, through their intentional and reckless/negligent release of the swarm, were the proximate cause of the foreseeable physical contact between the swarm and plaintiffs.

False Imprisonment

17. Defendants, through their intentional and reckless/negligent release of the swarm, were the proximate cause of numerous acts of confinement which kept Plaintiffs restrained.

18.

Trespass to Chattels/Conversion

19. Defendants, through their intentional and reckless/negligent release of the swarm, exercised dominion and control over the personal property of Plaintiffs, which was the proximate cause of partial or total damage to said property.

Negligent Infliction of Emotional Distress

20. Defendants, through their intentional and reckless/negligent release of the swarm, were the proximate cause of emotional and/or physical distress of the Plaintiffs.

Damages

21. The Plaintiffs seek recovery of personal injury damages including, but not limited to, pain and suffering, past and future medical expenses, and mental anguish.

22. Plaintiffs seek recovery for consequential damages and economic damages including, but not limited to, loss of consortium, conscious suffering, past and future medical expenses, and lost earnings.

23. Plaintiffs seek nominal damages.

24. Plaintiffs request an award of their attorneys' fees.

25. Plaintiffs seek pre and post-judgment interest and court costs.

26. Plaintiffs seek treble consequential and economic damages for Defendants' willful and wanton breach of duty owed to Plaintiffs and nonconformance with Federal regulations.

27. Plaintiffs seek treble mental anguish damages for Defendants' willful and wanton breach of duty owed to Plaintiffs and nonconformance with federal regulations.

27

28. Plaintiffs seek enforcement of Defendants' conformance with applicable laws and regulations under the Administrative Procedure Act.

Punitive damages

29. Defendants' actions were intentional, reckless, and/or grossly negligent and Plaintiffs seek exemplary damages on its strict liability and toxic tort claims.

30. Defendants' conduct is evidence of willful, wanton, or reckless/negligent disregard of a known and serious danger posed by release of unknown, self-replicating, and evolving nano-robots into the environment.

31. Defendants should be assessed punitive damages in an amount to be determined by the jury in order to discourage future wrongful conduct of a similar nature.

Jury Demand

32. Plaintiffs demand a jury trial.

Prayer for Relief

33. Plaintiffs respectfully request judgment against Defendants for the damages listed above, attorney's fees, costs, pre and post-judgment interest, and any other relief in law or equity to which Plaintiffs may be entitled.

Respectfully submitted,

[law firm and address]

By:_____

[attorney name]
State Bar No. _____

[attorney name]
State Bar No. _____

[attorney name]
State Bar No. _____

IN THE UNITED STATES DISTRICT
COURT FOR THE DISTRICT OF
NEVADA

JACK FORMAN, MAE CHANG,   )
ESTATESOF CHARLEY DAVENPORT,   )
VINCE REYNOLDS, JULIA FORMAN,   )
NATURALIST IN PARK, AND   )
PETER MORRIS   )
)   **No. 01-032009**
VS.   )
)
XYMOS, INC., LARRY HANDLER, CEO)
AND U.S. DEPARTMENT OF DEFENSE)

DEFENDANT'S ANSWER

Defendants, Xymos, Inc., Larry Handler, and the United States Department of Defense, do hereby deny all allegations made by the Plaintiffs, not herein expressly admitted, as permitted by Rule 8(b), Fed. R. Civ. P.

Defendant lacks knowledge or information sufficient to form a belief as to the truth of the allegations contained in Plaintiff's Complaint, except as expressly set forth below.

Defendants, by and through their attorneys of record, _____, of the law firm of _____, and for its Answer to Plaintiff's Complaint on file herein, admits, denies and alleges:

Jurisdiction

The district court has jurisdiction of the case that is docketed as No. 01-123123 pursuant to 28 U.S.C. § 1331.

General Allegations

1.   No response is required.

2.   No response is required.

3.   This paragraph states conclusions of law to which no response is required. To the extent a response is required, those averments are denied.

4.   The Defendants are without knowledge or information sufficient to form a belief as to the truth of those averments.

5.   This paragraph states conclusions of law to which no response is required. To

the extent a response is required, those averments are denied.

6. No response is required

7. The Plaintiffs failure to file an administrative claim bars their complaint against the U.S. Department of Defense under 28 U.S.C. § 2675(a). The Plaintiffs failure to state a "sum certain" for a particular claim precludes recovery on that claim. The U.S. Department of Defense should be dismissed from this complaint.

Causes of Action

8a. Plaintiff does not establish proximate cause. It is unforeseeable that the alleged breach caused the injuries.

8b. Plaintiff does not cite a statute that was violated; therefore, negligence per se is not applicable.

9.-10. Plaintiff does not define the alleged toxin. Plaintiff does not assert the level of exposure to the toxin. Plaintiff does not establish proximate cause for the injury.

11. These averments are denied.

12. Plaintiff does not establish proximate cause. It is unforeseeable that the alleged breach caused the injuries. Furthermore, there is no evidence that anyone has died. Most importantly, the alleged release of the swarm was neither the direct nor the proximate cause of the alleged deaths. The Plaintiffs own action was a direct intervening cause of death.

13.-14. The named Plaintiffs also engaged in the design and manufacture of the swarm which at the time of the design and manufacture was reasonably safe, was not defective, and was fit for the purpose intended. No warnings were necessary because the risk was not foreseeable. Most importantly, there is no proof that a defect existed when the product left the Defendants' control.

15. These averments are denied.

16.-18. There was no substantial certainty that harm would result from the alleged release of the swarm. There is no proximate cause element to intentional torts. The physical contact was not foreseeable.

19. Plaintiff does not establish proximate cause. It is unforeseeable that the alleged breach caused the emotional distress.

Affirmative Defenses

20.   Personal jurisdiction is lacking. Fed. R. Civ. P. 12(b)(2).

21.   Subject matter jurisdiction is lacking. Fed. R. Civ. P. 12(b)(1).

22.   Plaintiffs have failed to state a claim upon which relief can be granted.

23.   Plaintiff's injuries and damages, if any, were directly and proximately caused by the negligence, carelessness or fault of Plaintiff.

24.   Plaintiffs voluntarily assumed the risks surrounding the Xymos facility because they had actual knowledge of the particular risk, appreciated the magnitude of the risk, and voluntarily chose to encounter the risk

Wherefore, the Defendants requests that Plaintiff's claims against them be dismissed with prejudice and that the defendants be granted such other and further relief as the Court deems just and proper.

Respectfully submitted,
[law firm and address]

By:_____

[attorney name]
State Bar No. _____

[attorney name]
State Bar No. _____

[attorney name]
State Bar No. _____

XI.   NOTES on the Testimony of Jack Foreman

These notes are from quotes in the book, Prey, and should be included if you have Jack Foreman as a fact witness. Objections might include that some of his statements are hearsay.

RE: Biosafety and Biosecurity Regulations

Vince Reynolds (deceased), the maintenance director to Jack Foreman

"But these guys working here? Let me tell you, they're all fucking crazy. They're making these teeny-weeny little things you can see, pushing around molecules and shit, and sticking 'em togther. It's real tense and detailed work, and it makes them crazy. Every fucking one of'em. Nutty as loons. Come this way."
[Prey, p. 154]

The biotechnology design and process.
Wind, is the problem.
[Prey, pp. 190-192]

RE: OSHA and worker safety

"They're waiting for you, so let's get started. We got to take precautions, because this is an HMF, that's high magnetic field environment, greater than 33 Tesla, so . . ."
[Prey, p. 152]

RE: Risk and Environmental impact

Ricky Morse (deceased) said:
I [Jack Foreman] said, "Ricky, so what?" Those particles were scattered by the wind across hundreds of miles of desert. They'll decay from sunlight and cosmic radiation. They'll break up, decompose. In a few hours or days, they're gone. Right?"

Ricky shrugged. "Actually, Jack, that's not what ---"

It was at that moment that the alarm went off.
[Prey, p. 195]

RE: first sighting and description of a swarm
[Prey, p. 196]

Aren't there pictures? I think the environmentalists took pictures.

Well who cares? What will the pictures show, a dead coyote?.
[Prey, p. 144]

Apparently, the plant released some material into the environment. It was an accident. Now several dead animals have been found out in the desert. In the vicinity of the plant. [growling voice]

"Who found them?" Growly.

"Couple of nosy environmentalists. They ignored the keep-out signs, snooped around the plant. They've complained to the company and are demanding to inspect the plant."

"Which we can't allow."

"no, no."

"How do we handle this?" said a timid voice.
[Prey, p. 143]

"I say we minimize the amount of contamination released, and give data that show no untoward consequence is possible." Organized voice.

"Hell, I wouldn't play it that way," said growling voice. "We're better off flatly denying it. Nothing was released. I mean, what's the evidence anything was released?"

"Well, the dead animals. A coyote, some desert rats. Maybe a few birds."

"Hell, animals die in nature all the time. I mean, remember the business about those slashed cows? It was supposed to be aliens from UFOs that were slashing the cows. Finally turned out the cows were dying of natural causes, and it was decomposing gas in the carcasses that split them open. Remember that?"

"Vaguely."

"I'm not sure we can just deny --- "

"F**k yes, deny." [Prey, p. 144]

"Why the hell did they put it in Nevada, anyway?" [comment from Jack Foreman]

"Fewer regulations, easier inspections. These days California is sticky about new industry. There was going to be a year's delay just for environmental-impact statements. And a far more difficult permitting process. So they came here. [response from Growly]

Growly looked out the window at the desert. "What a shithole," he said. "I don't give a fuck what goes on here, it's not a problem." He turned to me [Jack Foreman] . . .
[Prey, p. 147]

"Those environmentalists are fucked," said Growly, with satisfaction. "They do medical research here, right?" [said to Helicopter pilot and Jack Foreman] [Prey, p. 147]

NO-C Environmentalists, Inc.
Photograph taken by John Nos, 500 yards from the Xymos building
Date:

Exhibit No. _____

RE: Medical Research and Biotechnology

"They do medical research here, right?" [Growly][Prey, p. 147]

[Vince Reynolds] "Yeah. Got a nest of rats in the building. Buggers kept getting fired. Literally. I hate the small of burning rat fur, don't you?" [Prey, p. 155]

"How did the rats get in?"

"Up through the toilet bowl." . . . "We've had a few accidents like that since I've been here."

"Is that right? What kind of accidents?"

He shrugged [Vince Reynolds]. "They tried to make these buildings perfect," he said. "Because they're working with such small-size things. But its not a perfect world. Jack. Never has been. Never will be."

I said again, "What kind of accidents?" [Prey, p. 155]

[Ricky Morse] Those kettles in the next room were indeed tanks for controlled microbial growth. But Ricky wasn't making beer--- he was making microbes, and I had no doubt about the reasons why. Unable to construct genuine nanoassemblers, Xymos was using bacteria to crank out their molecules. This was genetic engineering, not nanotechnology. [Prey, p. 187]

Jack Foreman's conversation with Physician RE: Negligence

Camper in the Sierras

Five days ago. But it's a completely different situation. This case involved a forty-two-year-old naturalist sleeping out in the Sierras, some wildflower expert. There was a particular kind of flower or something. Anyway, he was hospitalized in Sacramento. And he had the same clinical course as your daughter --- sudden unexplained onset, no fever, painful erythematous reaction."

"And an MRI stopped it?"

"I don't know if he had an MRI," he said. "But apparently this syndrome --- whatever it is --- is self-limited. Very sudden onset, and very abrupt termination." [Doctor]

"He's okay now? The naturalist?" [Foreman] [Prey, p. 67]

 "He's fine. A couple of days of bruising, and nothing more."

"Good," I said. "I'm glad to hear it." [Prey, p. 68]

XII. DEPOSITION EXCERPTS OF THE EXPERTS

Deposition of Expert for the Plaintiffs
The Expert is testifying on the issue of whether the defendant's activity was an abnormally dangerous one under Nevada law.

Dr. Roden

Attorney: Dr. Roden, can you tell us about the possible risk of the nanobots that were released into the environment by Larry Handler and Xymos?

Dr. Roden: Absolutely. The nanobots are self-assemblers, and the National Academies defines these self-assemblers as follows:

Self-assembly as described by the National Academy of Sciences report on the review of the NNI, is *the natural tendency of physical systems to exchange energy with their surroundings and assume patterns or structures of reduced free energy. Tandom thermal motions bring constituent particles together in various configurations, so that stable configurations (those with significant binding energy) form, tend to persist and eventually become predominant. . . information on how to assemble the structure is embodied in the structures of the individual components.*[1]

Attorney: But can you tell us if there is a risk in the release of these nanobots that are self-assemblers into the environment?

Dr. Roden: Absolutely. The self-assemblers pose a serious risk to the environment, and to the very existence of animals and humans, if the self-assemblers tend to organize effectively.

Attorney: But can you put that into some quantifiable measurement? What are the chances that releasing these self-assemblers into the environment would be harmful to humans?

Dr. Roden: Absolutely. I would say, about a 75% chance.

Attorney: A 75% chance that they would cause harm to humans if released into the environment?

Dr. Roden: Yes.

---

[1] National Research Council, National Academies, "A Matter of Size: Triennial Review of the National Nanotechnology Initiative," The National Academies Press 100 (2006).

Curriculum Vitae

Dr. Roden

444 Las Vegas, Nevada, 89104 – (888) 434-4444 –
dr.roden@hotmail.com

Experience

Director

HighTech, Las Vegas, Nevada

2002 – Present

Supervise the research and development of molecular nanotechnology.

President

NanoSmart, Inc., Las Vegas, Nevada

1993 – 2002

Responsible for the management and growth of NanoSmart, Inc. In addition to managing all the operations, he is the primary consultant to clients on all nanotechnology regulatory, environmental and health issues.

Managing Senior Engineer

Bots Corporation, Las Vegas, Nevada

1985-1993

Was the project manager for a variety of environmental projects involving evaluation and assessment of environmental regulations.

Assistant Engineer

Fright, Inc., Las Vegas Nevada

1973-1985

Assisted the director in a variety of projects involving the research and attempt of self-replication in nanotechnology.

Education

Masters in Biological Engineering

Harvard, Cambridge, Massachusetts

May 1975

Bachelor of Science in Chemical Engineering

Harvard. Cambridge, Massachusetts

May 1973

Publications

Nanotechnology Journal

Deposition of Expert for the Defendants

The Expert is testifying on the issue of whether the defendant's activity was an abnormally dangerous one under Nevada law.

Attorney: Dr. Lindman, can you please tell us about the risk of nanobot swarms which are claimed by the plaintiffs to be responsible for damages to the environment and the deaths of several people?

Dr. Lindman: While conceivable, it is far from reality to devise nanomachines with self-assembly capability, with ultimate potential to consume all life is what John Tierney wrote in an Op-Ed., entitled, Homo Sapiens 2.0, N.Y. Times, Sept. 27, 2005, at A25 (summarizing an apocalyptic scenario resulting from out-of-control nanobots).

Attorney: So are you saying that the nanobots are actually not possible?

Dr. Lindman: I am saying that "Grey goo" is the term sometimes used to refer to the material resulting from self-replication run amok. A term that was first mentioned by Drexler. Drexler said that "there is no reason to build anything remotely like a 'grey goo' device" and that more probable scenarios such as nanotechnology-based weaponry warrant more attention) and that ("[T]he easiest and most efficient [molecular manufacturing] systems will not have the capabilities required for autonomous runaway manufacturing.").

XIII. EXHIBITS FILE

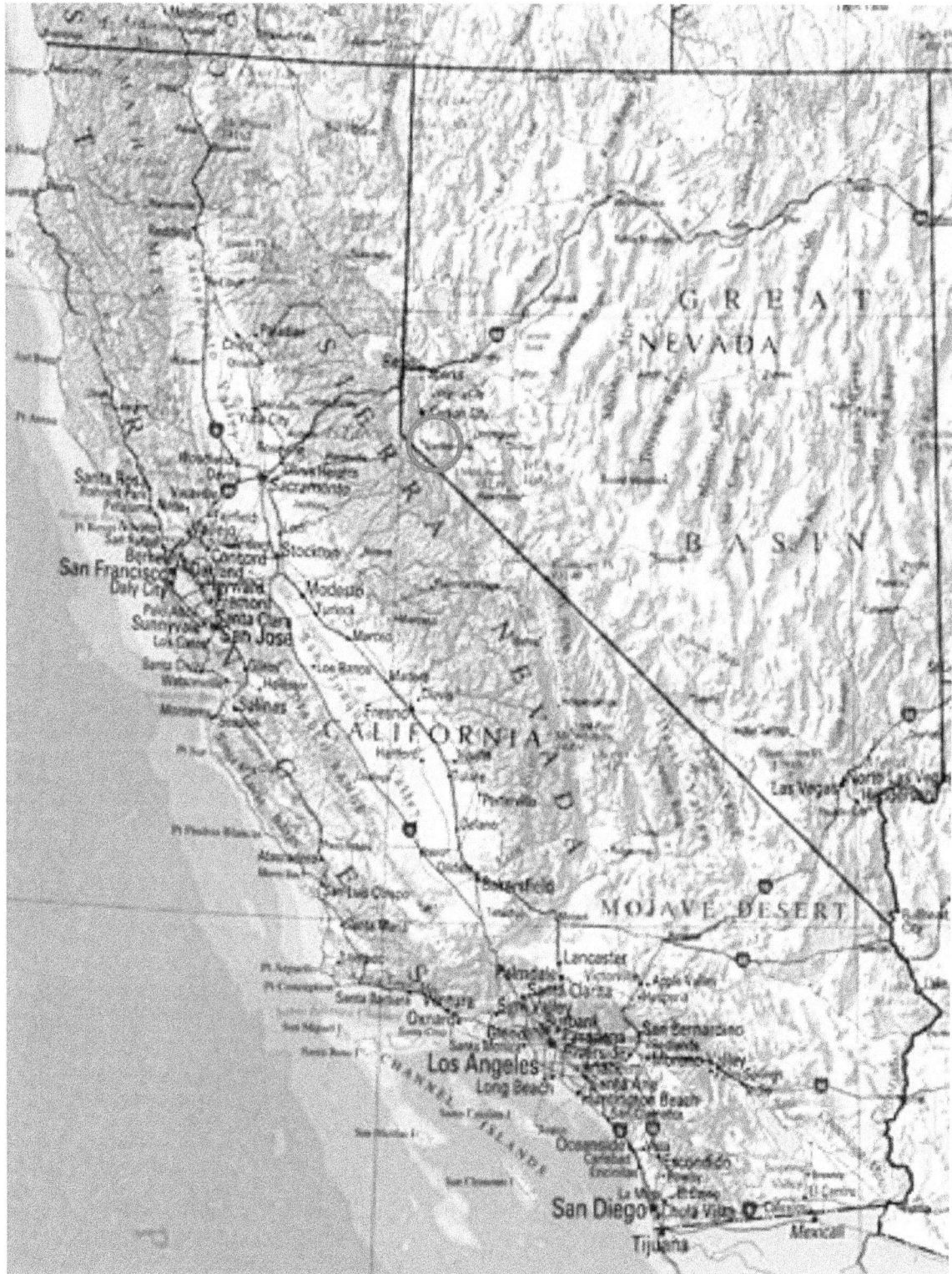

Pilot: "We just crossed the Nevada line. Another ten minutes." *[Prey, p. 145]*

Reference article

Don't let Crichton's Prey scare you--the science isn't real

Chris Phoenix

January 2003

(Chris Phoenix is co-founder of the Center for Responsible Nanotechnology)
In order to explain what's wrong with the science in *Prey*, this review contains spoilers.

Imagine a horror story about baseball, in which the batter keeps hitting the ball hard enough to kill the fans. The story might be entertaining, but it's obviously unrealistic. Suppose further that at one point in the story, the author writes about someone walking back to the dugout after three "fouls." Does the author not know the difference between a foul and a strike, or was he simply in too much of a hurry to bother getting the words right? Either way, no one could learn the rules of baseball from that story. Even if it was mostly right, a few wrong facts make all the difference--especially if the reader does not know which facts are wrong.

Prey contains comparable exaggerations and mistakes in science. A scanning probe microscope and an electron microscope are basic tools of nanotechnology, and they're not even remotely similar. Yet Crichton confuses the two, on page 133. He also confuses piezoelectric with photovoltaic. And he writes about a nanobot that's "one ten-billionth of an inch in length." This is the size of a single atom, not a whole robot. But these are simple errors; the flaws in his nanotech run deeper.

Crichton's hypothetical nanotechnology is built in a multi-stage process. First, bacteria produce chemicals, which are modified and then combined to form "assemblers." The assemblers are attached to other bacteria, which produce more chemicals that are combined by the assemblers into the final product: a tiny, flying robot with an onboard computer, solar cell, and other useful gizmos. Remember that the bacteria are attached to the assemblers, not to the final swarm-bot product. And the bacteria are not involved in the function of the final swarm-bot (nor is the swarm-bot involved in the assembly process). But for some unexplained reason, people infected with the swarm-bots melt (in seconds) when splashed with a bacteria-killing virus, which should have no effect on the swarm-bots. It makes no sense--it was simply necessary to the story.

Even this is not the biggest problem with the science. The cornerstone of the book is evolutionary learning and emergent behavior. Crichton's explanations are too superficial to allow detailed criticism, but there are a few obvious impossibilities. For example, the swarm supposedly "learns" by reproducing itself: the dumb swarm-bots die off, and the more effective ones survive to reproduce. (How their program is fed back into the assemblers is not explained.) But when the swarm first infects the humans, it immediately begins to coexist--and even to make them look healthier, and to modify their behavior in subtle ways. The swarm can't reproduce in humans. For one thing, there's no gallium or arsenic for the electronics. (There's not much gallium or arsenic lying around a desert, either.) So how did the swarm learn how to get along with human biology and cognition? There's no way it could have.

49

Crichton's introduction, "Artificial Evolution in the Twenty-first Century," appears to be a serious attempt to warn us about the dangers of technology that is capable of evolution. He closes by saying that if someone manages to create evolving organisms before they can be regulated, "... it is difficult to anticipate what the consequences might be. That is the subject of the present novel." The scientific explanations scattered throughout the book increase the apparent plausibility of the story. Even the name Xymos is a clear reference to the real-world nanotech company Zyvex. It seems that Crichton is doing his best to scare the readers about real-world nanotech--he wants his audience to believe that the scenario in the novel could actually happen as described! He might succeed in scaring people; a friend of mine who's a geneticist told me that Jurassic Park set back public perception of genetic engineering by a decade. This would be unfortunate, because the Prey scenario contains so many implausibilities--and impossibilities--that in the end, the reader will have learned nothing about the actual risks of nanotech.

It is worth listing a few more of the exaggerations in the book. For example, Crichton describes glass as being unsuitable for handling chemicals: "At the molecular level, glass is like Swiss cheese, full of holes. And of course it's a liquid, so atoms just pass right through it." (p. 131) A little common sense shows that this is, at best, stretching the truth. Light bulbs, vacuum thermos bottles, and TV tubes can last for decades; they all depend on the fact that atoms normally can't pass through glass. (Helium can make it through--in vanishingly small amounts--but helium is completely inert.) Another exaggeration is the idea that the swarms could coevolve with worms in a matter of days. Even if the swarms could evolve that quickly, the worms could not. Finally, a significant impossibility is the idea that a swarm-bot could fit inside a synapse (p. 256). A synapse is only a few atoms wide; the swarm-bots are hundreds or thousands of times bigger. But without this impossibility, a major sub-plot falls apart.

The exaggerations are not only about nanotech. Near the end of the story, a character sets off all the fire sprinklers by melting one of them. Despite what we've all seen in the movies, sprinkler systems don't really work that way. Crichton's characters also explode welder's thermite to destroy the clouds of nanobots. According to the welding supply company I called, thermite is used for welding because it burns hot enough to melt metal--but it does not explode. I was told it probably wouldn't even ignite a sheet of paper a few feet away.

This is not to say that nanotech is completely safe. The possibility of "gray goo"--self-replicating nanobots--has been discussed from the beginning. But gray goo would be very difficult to design. It would be far more complex than a car--probably more complex than the Space Shuttle. General Motors recently made headlines by taking only a few months to design a car. It's completely implausible that a failing company could create an evolving gray goo by re-engineering a specialized product in a matter of weeks; this same company couldn't even solve the relatively simple problem of keeping the swarm together in a breeze. Remember that the swarm-bots don't directly replicate; they are built by assemblers using bacterial chemicals. Among other tasks, the scientists would have had to rapidly invent a way to transfer the evolved program out of the successful swarm-bots and feed it back into the assemblers or the bacteria to produce the next generation. This would require a completely new set of molecular machinery.

In Crichton's stories, the scientists are mad--all but one who moans about how "nature will find a way" and "we should not play with things we don't understand". In the real world, it's the other way around. We won't have nanobots for years, maybe decades, but scientists have already written a code of practice, the "Foresight Guidelines on Molecular Nanotechnology," that would prohibit anything remotely like what Crichton has invented. And because evolution doesn't work as magically as Crichton portrays it, scientists probably wouldn't even be tempted to release evolving nanobots; in real life, the swarm would have been destroyed almost immediately, and so would never have had a chance to improve itself. (You can find the Foresight Guidelines at http://www.foresight.org/guidelines/current.html)

Nanotechnology is genuinely interesting, and in some ways even scary. Hopefully, Prey will generate interest in the subject; reliable information about nanotechnology is available on-line and in numerous books. Readers of Prey should remember that it does not provide a realistic portrayal of the technology. Many parts are impossible, and many others are stretched beyond plausibility. For real, scary science, read non-fiction such as The Demon in the Freezer by Richard Preston. Smallpox isn't as trendy as nanobots--but unlike nanobots, smallpox already exists, and could easily be used by terrorists. Prey is not scary science because (to be blunt) very little of it is real; those who let themselves be scared by it might as well wear protective gear to the next baseball game.

Author's note: This review replaces an earlier version that was overly negative. The author believes that information should not disappear from the Internet, and has asked the editors of this web site to retain the original version [here]. The facts presented in the original are essentially the same; the author retracts the style.

Chris' interest in nanotechnology began when he took Eric Drexler's class "Nanotechnology and Exploratory Engineering" at Stanford University in 1988. He has followed the field continuously since then, attending numerous nanotech conferences, contributing frequently to several on-line discussion lists, and helping to review a major book and a Ph.D. thesis on nanotech. He is a Senior Associate of the Foresight Institute and co-moderator of the sci.nanotech newsgroup.

Other papers by Chris

See also these other articles on Prey and Nanotechnology:

 Michael Crichton is very, very afraid of technological progress—again.

Reason.com December 11, 2002 Are Crichton's horrific fantasies based in reality? What evidence is there that humanity rushes headlong into misusing powerful new technologies? Practically none. Instead of using computerized probes for mind control, physicians implant them to control Parkinson's disease. Instead of carelessly bringing space viruses to Earth, NASA set up elaborate containment and decontamination systems for astronauts returning from the moon and any future remote explorers bringing back samples from other planets...

Falling Prey to Science Fiction "...so I'm going to point out something almost as obvious about Crichton's book: the factual situation that he relies on for his story is one that could only happen if the researchers in question were (1) stupid; (2) criminally negligent; and (3) willing to violate the consensus ideas about nanotechnology safety." Glenn Harlan Reynolds on Michael Crichton's new novel Prey.

Foresight Guidelines on Molecular Nanotechnology | The future dances on a pin's head | Forward to the Future: Nanotechnology and Regulatory Policy | Small is evil | Responsible Nanotechnology: Looking Beyond the Good News | the complete review | Trouble in nanoland

Opinions stated are the author's, and do not necessarily reflect those held by 7thWave, Inc.

Chris Phoenix, "Don't let Crichton's Prey scare you --- the science isn't real," Center for Responsible Nanotechnology, 7thWave, Inc. (Jan 2003) at http://www.nanotech-now.com/Chris-Phoenix/prey-critique.htm .

4 Nanotechnology L. & Bus. 189, Nanotechnology Law & Business (June 2007)
Legislative & Regulatory Affairs
EXCERPT

NANOASSEMBLERS: A LIKELY THREAT?
Martin Moskovits [FNa1]

ABSTRACT
In his popular book Engines of Creation, author and innovator, Eric Drexler, proposed the concept of a nanoassembler, a tiny robot equipped with the ability to construct with the aid of many others, useful objects by identifying raw materials at the atomic, molecular or, at least, the nanometer scale, then assembling these tiny constituents into a complex structure. The nanoassembler would also have the ability to reproduce itself in anticipation of a given task, then, presumably, disassemble those nanoassemblers no longer needed, harvesting the raw materials for subsequent projects. Several individuals (including Drexler himself) have expressed concern about the prospect of the dust-particle-sized nanoassemblers replicating themselves uncontrollably, leading to large parts of the earth's surface being covered in a blanket of nanoassembler-dust, a material christened "gray goo" by Drexler. A great deal of concern has been expressed regarding personal and environmental risk posed by technological developments in nanotechnology. The threat posed by the possible development of nanoassemblers depends, of course, on the range of capabilities that such nanoassemblers could possess. Some have suggested that in carrying out its prescribed duties, a nanoassembler, in the full incarnation proposed by Drexler, would likely have to contravene either or both the second law of thermodynamics and the uncertainty principle of quantum mechanics. In this article, Dr. Martin Moskovits assesses the likelihood of creating nanoassemblers of the type proposed by Drexler as highly unlikely. This does not preclude the development of useful microscopic robots with far more limited abilities than those proposed by Drexler, such as a micro-robot that navigates the circulatory system diagnostically. These, however, would not need to be self-replicating and would therefore not pose the same level of risk as Drexler's nanoassemblers. Simple, microscopic, self-replicating mechanical systems would also be possible, which could pose health and environmental threats not unlike pathogens, but a nanoassembler that carries out wholesale molecular assembly intelligently and cooperatively as proposed by Drexler is unlikely.

I.   INTRODUCTION

Nanoscience & technology ("nanotech") is a field of research, engineering and business that is currently being pursued by thousands of individuals and entities. Despite its allure to many, there is no universally accepted definition of Nanotech. No single, easily expressible unifying mission exists, other than manipulating matter at the nanometer scale. Nor is there a single revolutionary technology that defines nanotech, unlike, for example, biotechnology which was based on the gene-engineering capabilities made possible by the development of recombinant DNA. Additionally, two versions of nanotechnology exist, at least in the mind of the public: the version practiced by most *190 scientists and engineers [FN1], and the one popularized by Eric Drexler. [FN2] Most scientists and engineers do not currently work towards fulfilling the latter's vision.

Nevertheless, Drexler's picture of the fulfillment of nanotech's future promise still lingers in the public's mind. Some, understandably, dread the possibility that nanotech will lead to a world overwhelmed by self-replicating sub-microscopic robots who, like the sorcerer's apprentice, once started cannot be stopped, resulting in what Drexler called "gray goo." This scenario and its sci-fi progeny, as described, for example, in Michael Crichton's book Prey [FN3], has produced some sensitivities and fears. I should state that recent, legitimate concerns raised by scientists and members of state and federal legislative bodies regarding the health and environmental impact of nanotechnology are not based on the possibility of gray goo. This is because the vast majority of those who work in nanotech are not pursuing research along the lines advocated by Drexler. [FN4] Of greater concern to researchers are issues of human health and environmental dangers posed by both the synthesis of novel materials and by the fabrication of well-known materials but in hitherto-unachievable submicroscopic forms. [FN5] One can get a handle of these dangers, however, since they are not greatly dissimilar from those posed by research in chemistry, biochemistry, and biology. Similar to these other disciplines, the costs of training and implementing human and environmental safety programs in nanotech are a component cost of carrying out research responsibly. This aspect of nanotech safety is currently a significant area of interest for both researchers and the media. [FN6] In this article I will focus almost exclusively on the prospects of nanoassemblers.

II. NANOTECH RESEARCH: BRIEF HISTORY TO THE PRESENT

A word about the version of nanotech carried out by the vast majority of its practitioners is warranted. Nanotech is an outgrowth of the activities of the 70s and 80s in fields such as cluster science, [FN7] which emphasized "scale-dependent" or "size-dependent" properties of matter and of technological advances in microfabrication. Questions such as how bulk properties arise when an atomic cluster consisting of, for example, a bonded assembly of silicon atoms is progressively made larger was a typical consideration in "cluster science." This field became rather vibrant with many researchers carrying out experiments on mass-selected clusters produced in molecular beam machines, [FN8] leading to, for example, the discovery of C60. The many insights that came out of cluster science included an understanding of the development of bulk electronic, magnetic and optical properties of metal or semiconductor clusters as they progressively became larger as well as the very complex chemical properties of clusters as a function of cluster size. One outcome was the realization that bulk-like properties arose at different size scales depending on the property being investigated.

*191 Many of the fundamental questions being addressed by nanotech have even earlier roots. Basic questions regarding size-dependent properties of matter were being asked in the heyday of colloid chemistry. In the early years of the 20th century, physical chemistry was almost synonymous with colloid chemistry. [FN9] The luster of colloid science dimmed largely because scientists were able to ask deeper and more probing questions than the instrumental and computational arsenal of the day could answer. Many of these questions were resurrected in the 1980s and 1990s and are now being addressed in the context of nanotech through today's powerful technological and computational tools.

Cluster science was only one of several antecedent fields that gave rise to nanotech. Advances in mircrofabrication [FN10] such as e-beam lithography, in synthetic chemistry, polymer science [FN11] and supramolecular chemistry [FN12], in theory [FN13] and in computing, also paved the way to nanotech. Finally, intangibles, such as a greater propensity for university researchers to work as part of large collaborations involving scientists from a variety of disciplines, made nanotech and its inherent interdisciplinary approach a possibility. This mini-cultural transformation came about as a result of a whole host of little things such as the development within universities of mechanisms for assessing success of individuals participating in collaborative research, the availability of research funding earmarked for group research activities and Centers, and the realization that some of the major technical problems of the present day require the talents of a broad range of disciplines for their solutions. Not all of these collaborative efforts have been successful, of course, and the ease with which interdisciplinary research can flourish varies from institution to institution even today, but a significant enough fraction of such cultural changes have succeeded to make the phenomenon "real."

Questions rooted in scale- and size-dependent issues at the nanoscale go further back still. Michael Faraday's sample of colloidal gold is still on display in the Royal Institution. (Faraday understood what he had and had correctly estimated the colloidal particle's size.) We all know Feynman's famous presentation, "There's Plenty of Room At The Bottom," in which he enunciates a couple of challenges in miniaturization. [FN14] Photography was known to depend critically on the formation of small silver particles - the so-called latent image - whose properties depend critically on silver cluster size. [FN15] Similarly, in heterogenous catalysis many catalytic processes depend strongly on the nanoparticles' size, geometry, and other structural properties. [FN16] The observation, approximately 25 years ago, of so-called surface-enhanced Raman spectroscopy was recognized early on to be a genuine nano-phenomenon in the sense that the effect is observed only with metal systems fabricated with nano-scaled constituent parts. [FN17]

Most scientists working in nanoscience have avoided working on creating nanoassemblers, and do not subscribe to Drexler's definition of nanotech in terms of the nanoassembler. Instead, they work on creating nanomaterials, nanofabrication, micro- and nanoelectromechanical systems, nanoimaging *192 technologies (e.g., scanning probe microscopies), nanoelectronics and photonics, bio-nano (e.g., nano-based drug delivery), sensing and diagnostics (e.g., chip-based diagnostics), biomaterials and prosthetics, nano-inspired biotechnologies, bio-mimetics and bionics, and the like. [FN18] Researchers have had a number of early successes both in terms of product and process. For example, the use of modified ink-jet printing as a manufacturing technique for making products such as flat displays has potentially dramatic value. Additionally, the cosmetics industry has been significantly transformed through its use of nanomaterials.

III. NANOASSEMBLERS

In view of its strongly evocative role in the field, it is worth considering the prospects for creating a nanosassembler. In creating such an entity, researchers need to agree on a specifications sheet. The nanoassembler, or nanobot, has gone through a series of evolutionary modifications since its original suggestion, largely in response to skepticism from mainstream scientists and engineers. The neglect of the nanobot by the nanotech community has itself elicited a broad range of reactions. Drexler, for example wrote an opinion piece in the New York Times [FN19] regretting the fact that most research in nanotech is not directed towards developing nanoassemblers. There was also a rather public debate between the late Nobelist, Rick Smalley, and Drexler on the feasibility of nanobots. [FN20] Supporters have responded to skeptics by pointing to viruses as prototypes of self-replicating nanobots. In my mind there is a vast difference between the spec-sheet proposed for the nanobot and what a virus is asked to do, which is essentially only replicate itself.

As a result of this counterpoint the concept of the nanoassembler has evolved since its original introduction, generally in the direction of a reduction in its capabilities. [FN21] With enough conceptual evolution and a greatly reduced range of specifications, the nanoassembler could ultimately become something tractable such as a microscopic robot that can perform a number of useful functions; perhaps it could even replicate. But even if the nanoassembler cannot replicate, the notion of an ultra-small robot that can, for example, navigate the bloodstream performing microsurgery or activating neurons so as to restore muscular activity, is not an unreasonable goal, and one that may be realized in the near future.

The issue that the science and engineering community has with the nanobot is not primarily associated with the societal alarm that a self-replicating intelligent and purposeful dust particle would pose (e.g., evolutionary possibilities that could cause it to reprogram itself to reproduce without limits or attack its human creators). Rather the community is skeptical because of the fundamental scientific barriers limiting the scope of this idea.

In its original form, the nanoassembler is a small creature that can be programmed to team up with others of its kind to construct a complex structure from materials it seeks and retrieves from its environment. The nanoassembler can self-replicate so that when facing a major construction project, its first order of business is to build the team. Near the end of the project it might deconstruct the team for its material content (or keep them on for other projects). It might be reprogrammed wirelessly for each set of tasks or even various stages of a given task. The nanoassembler can make intelligent decisions *193 based on its database, wirelessly access information using the Internet, or even create a private network of its own data purposely kept out of the reach of humans. The on-board program might specify a declarative goal (e.g. create one million bushels of wheat kernels from the contents of a land-fill site") subject to a few helpful restrictions (e.g., do not kill any people or destroy useful articles such as homes and cars). The rest is downloaded from databases. This is an attractive, ultimately utopian, goal that would free the human species from all of its problems - one specifies the goals: feed, house and provide care to all humankind, and the army of nanobots both develops the strategies and fulfils them.

To be truly useful and revolutionary the nanobots need to be small enough to assemble their structures one or a few atoms or molecules at a time. Protein for food, for example, might be assembled from amino acids retrieved from sewage. An entire book might be manufactured from sawdust and a handful of other ingredients, while its printing is meticulously set in place one nanoscopic carbon dot at a time.

I will now argue that the nanoassembler in this original form is, in fact, a Maxwell Demon. Nineteenth century physicist James Clerk Maxwell developed a thought-experiment to illustrate the workings of the second law of thermodynamics in which he proposed a small intelligent creature that operated a very small trap door covering a very small hole in a wall separating two rooms. [FN22] The rooms are filled with a gas (e.g., air) at room temperature. The molecules of the gas move with a vast range of speeds that obey a distribution, called a Maxwell-Boltzmann distribution. A sizable fraction of the molecules travel with speeds far in excess of the average and another sizable fraction travel with speeds much less than the average. The Demon is appropriately situated to see all of the molecules on either side of the trap door. It is programmed in such a way that the Demon opens the door when it sees a slow molecule approaching the trap door from the right or a very fast molecule approaching the trap door from the left. After enough molecules are sorted by the Demon in this way, the room on the left becomes colder and the room on of the right becomes hotter because the Demon selectively directs molecules to the two rooms. (This is a nanoproject since the hole has to be sufficiently small to ensure that only the desired molecules pass when the trap door is opened.)

The difference in temperature that results from the Demon's energy-sorting action could be used to power an engine in the normal way. That is, energy flowing from a hot place to a cold place could drive some sort of mechanical device in the process. The Demon's actions plus the engine connected to the two rooms creates a device that sucks energy out of air at a constant temperature; cooling the air and extracting useful work. This process does not contravene the first law of thermodynamics since the work extracted (plus the work needed to operate the Demon) need not exceed the work extracted from the air. It does, however, contravene the second law of thermodynamics, because it essentially takes a system at a constant temperature and converts it into two systems: one hot and one cold. That is, the process decreases entropy without expending energy. Thermodynamics teaches that such a process contradicts the second law because the reverse process (i.e. heat flowing from hot to cold) is the spontaneous process. Thus, to the extent we believe the second law, we could never create such a Demon. The beauty of the Maxwell Demon is that it forces people to think microscopically about the second law of thermodynamics, which is a law governing systems of molecules (i.e. macroscopic assemblies rather than individual particles). It also illustrates that, while not purposely postulated to do so, plausible-sounding processes may nevertheless contravene the second law.

A great deal has been written about Maxwell's Demon, and why construction of one is impossible. One of the most beautiful and easy-to-understand expositions is given by Feynman in his famous physics *194 course. [FN23] There have been innumerable propositions made over the years to get around the limitations of the Maxwell Demon, just as there continue to be schemes proposed to construct perpetual motions machines of the second type (i.e. machines that do not create energy out of nothing but defy the second law of thermodynamics). Many of them try to get around the problem by confusing size scales in trying to make the Maxwell Demon large enough so as to make it less prone to thermal noise. In some discussions the Demon is replaced with a less obviously flippant embodiment, which might be regarded as prejudicial to an open-minded discussion. Feynman, for example, proposes a very light ratchet that can turn in one direction. [FN24] The ratchet is attached to a paddle-wheel which allows fast molecules approaching from one side to turn the wheel while those approaching from the other side cannot. By making the paddles small enough, few molecules strike the paddle at any given time. Thus, the wheel turns because the number of molecules striking the paddle from one side would often not balance out the number of molecules striking the paddle from the other side. Feynman explains that this arrangement does not work; when the system is made small enough to create the asymmetry in the number of molecules striking the tiny vanes, the thermal "chatter" in ratchet allows the paddle-wheel to turn in both directions, thereby destroying the one-way motion that would defy the second law of thermodynamics.

Actually, one would be perfectly happy if the nanoassembler could function as a Maxwell Demon (i.e., extract lots of energy from the atmosphere). Cooling down the atmosphere by as little as one degree Celcius would provide us with ~3 x 1021 kJ of energy corresponding to approximately 8-years' worth of global energy usage. Reducing the temperature of the atmosphere slightly is not a bad thing to do these days. Much of the extracted energy would be restored to the atmosphere through its use, thus this mechanism (which is impossible according to the second law of thermodynamics) would provide us with energy for a long time. With all of that cheap energy available the need for nanobots as a structural tool becomes less urgent.

Three reasons why Maxwell's Demons and nanoassemblers (in their original incarnation) are unlikely to be achievable:

1. Very small objects are composed of very few molecules and atoms. To pick up a specific molecule from the environment and place it precisely where you want it, you would need "limbs" that are not very much bigger than a molecule. Such limbs are themselves composed of relatively few atoms or molecules. As a result the limbs are subject to thermal fluctuations, Brownian motion, and a quantum phenomenon called zero point motion. Thus, the nanobot's limbs will jitter and their accuracy as an assembler will be faulty. A nanoassembler reaching for a given molecule will often miss, and because the jitter (both of the nanobot's limbs and the molecule) is unpredictable, its reliability would be compromised.

2. For a nanobot to pick up a desired molecule from the environment it will need to have on-board analysis capability. The analytical tool may be some form of spectroscopy that shoots photons towards the location of a putative molecule and then reads the returning photons. Quantum mechanics says there will be an uncertainty in position and momentum involved in this analysis process preventing the nanoassembler from knowing the molecule's location at any given time - another limitation on the precision of the nanoassembly process. The nanoassembler will, therefore, reach for the molecule it deems useful only to discover that the molecule is not where it thought it was, or worse still, replaced by something whose identity is unknown. The object "assembled" under

those circumstances, if successful, will not meet specifications: the food might be toxic, and the book's pages filled with gibberish.

3. There is a presumption in the nanoassembly process that molecules can be collected from the environment with little energy cost, and that these molecules will stay put when placed in the ideal spot by the nanoassembler. In fact, there are very few "free" molecules because most molecules comprise stable compounds or solids. To remove stable molecules from their matrix requires significant energy. Likewise, molecules often have barriers to bond formation. Thus, even if they are placed near the ideal location they will not necessarily form a bond, but will move to another location when released by the nanobot. This is why most natural assembly processes, such as those involved in living systems, do not use molecule-by-molecule assembly, preferring, instead, "self-assembly" as a structure-forming mechanism. (Self assembly is the process whereby chemical building blocks assemble into chemically possible structures spontaneously, driven by normal chemical principles and finding the correct binding sites after making many random attempts. This process works efficiently because molecules can generally make very many attempts per second. Both evolution and people have exploited the propensity for self assembly as a principle for rational design: the former by evolving the correct building block through natural selection over time; the latter by using the talents of a synthetic chemist to rationally design building blocks having a propensity to self-assemble into the desired structure.)

A mechanism often encountered in complex structures is hierarchical assembly. Muscle, for example, is assembled from molecules that first come together to make fibrils. The fibrils assemble further into superstructures (fasciculi), that, in turn, assemble into muscles. Each of these processes may use different chemical and organizational strategies and may opt for very different levels of fault-tolerance as the size scale changes. Most protein molecules, for example, have very precise atom-by-atom compositions and structures. Fibrils can afford to have somewhat less perfect structures and yet perform their tasks adequately. The manner in which they aggregate into muscle tissues could afford to be even less precise and more tolerant towards organizational errors (i.e,. the organization error as a fraction of the size scale of the structure in question).

Once again, a nanoassembler would be possible if its requirements are sufficiently undemanding. But, its "revolutionary" nature and the value added by the concept would then be greatly reduced as the difference between what it does and how that differs from existing technologies and life processes carried out, for example, by enzymes, genetically-modified single-cell organism and the like, becomes indistinct.

Aside from restrictions by physical laws, issues of efficiency also require consideration because they may, for example, translate into economic barriers. Nanoassembly of the sort discussed above is essentially building large objects with molecular precision right down to the molecular, or perhaps even the atomic, level. It also requires construction by adding material, with great precision, at the surface of the three-dimensional object. Such addition is generally an inefficient process because the number of molecules contained in even a relatively small object is immense. Let me illustrate. Assume we want to build a brick home. The bricks themselves are produced in a "self-assembly" process. That is, mortar, crushed stone, and other ingredients are mixed randomly and allowed to "set" in a mold. Within the brick there is little control as to which precise particles bind together. Also, parameters such as the mean-distance between particles are allowed to have a broad distribution of values with no precise control on the distance between any two specific particles. The specific arrangement of particles will vary from brick to brick and will be established by a globally self-organizing process that occurs throughout the brick as it undergoes the hardening process (which is a kind of complex mineral polymerization reaction). Although the precise locations of the particles and their interconnects are variable, certain statistical characteristics will be fairly constant and predictable from brick to brick for the bricks to have optimal properties. These characteristics will be determined by such parameters as the composition of the mortar from which the bricks are manufactured, the temperature and humidity at which they are baked, the time of curing and so on.

In contrast, arrangement of bricks in the house walls requires a greater level of precision relative to its size scale, but an even lower level of precision on the molecular size scale: the brick-to-brick distance may vary by billions of molecular diameters. But relative to its size scale (which is reckoned in centimeters) the distribution of the brick to brick distance will be less variable than the relative inter-particle distances of the particles composing the brick. Higher-level structural elements of the house might require even higher levels of precision (again relative to its size scale). The surface of the parquet floor, for example, will have to be very smooth, so as not to be rough on bare feet. The point is that for most large manufactured structures, we are able to a demand greater and greater level of relative precision at the higher dimensional scales (where it is easier to achieve) while reducing the precision relative to the molecular size scale. This leads to great efficiency.

If, on the other hand, we required molecular level precision for structure elements regardless of their size, efficiency and therefore affordability would be hard to maintain. Assume, for example, we want nanobots to assemble a 1 cm3 brick by laying down one molecular layer at a time over a 1 cm2 surface. To complete this tiny brick, the nanobots would have to lay down approximately 1×10^7 layers. If each layer could be laid down in 1s then it would take roughly 115 days to assemble the brick. Laying down layers one molecular level at a time is a highly optimistic estimate because nanobots would have to fetch molecules from a pile of raw materials which might require a (relatively) great journey. Additionally, nanobots would have to work in a rather cramped (i.e., 1 cm2 surface) environment in which millions of other nanobots have to place their loads accurately. Greater efficiency may be achieved if the nanoassemblers worked on all six sides of the cube rather than on one (although the rate of assembly would be improved by a factor of less than 6, since at the outset the surface area is less than 1 cm2 allowing a relatively few assemblers to work on the brick at the start of the process), but the process would still remain rather inefficient.

Layer-by-layer atomic and molecular fabrication is currently in use by the semiconductor industry in crystal growth and molecular beam epitaxy (MBE). [FN25] This approach is generally only efficient when the desired final product is relatively small and thin. Even here, a great deal is accomplished by self-assembly. When growing a film by MBE, for example, the atoms descend on the epitaxial substrate randomly and hunt for crystallographically ideal resting locations by surface diffusion. If instead we required a nanoassembler to walk over the surface and place the atoms at specific and predetermined sites, the construction of even a thin-film device would likely be inefficient.

Recently, the nanoassembler has undergone significant modifications. One such concept is the table-top nanoassembler which is a set of turning gears that deposit molecules, one at a time, in the right locations in a structure that, presumably, is positioned appropriately by a mechanism below the "gears". An artistic rendering was published on the cover of the December 1, 2003 issue of Chemical and Engineering News. [FN26] The commingling of size scales in that figure illustrates the propensity to confuse size scales when discussing the nanoassembler. The gears seem to be fabricated out of a metal composed of atoms that are infinitesimally small compared to the molecules they carry. The table-top nanoassemblers no longer have to scurry about finding the required raw materials which, to me, is a major blow to the concept. One of the most attractive aspects of the original idea was to have an army of nanobots permeate a garbage dump taking apart our refuse and reassembling it into useful products. With the spec sheet for the nanoassembler reduced to more tractable functions, it is no longer clear what these stripped-down versions of the nanoassembler bring to the table over and above existing nanofabrication technologies other than a utopian redolence of the original item.

Reducing the level of threat posed by the likelihood of creating gray goo does not mean there are no issues of personal and environmental health to worry about in nanotech. The major issue is the fact that large numbers of scientists and engineers are creating new materials, and known materials, at an almost unprecedented level of subdivision. These materials, both new and those known to be benign in the bulk state, should be (and are) considered toxic until shown otherwise. One cannot claim that new, useful, and, at times, surprising properties arise as a result of this nanoscale subdivision without, at the same time, worrying about the possibility that among those novel properties are highly undesirable, even dangerous ones. A subset of these new properties may include new photonic properties that photoexcite nanoparticles when illuminated at wavelengths at which their bulk counterparts are transparent. Fortunately these are currently robust areas of research. [FN27]

IV. . CONCLUSION

In summary, most researchers currently practicing in nanotech are not attempting to create nanobots. This is partly because, as originally described, nanobots would challenge the second law of thermodynamics in carrying out some of its prescribed duties. Molecule-by-molecule assembly is also shown to be an inefficient means for creating macroscopic objects. Accordingly, swarms of nanobots and gray goo is not a likely health and environmental threat. Most nanotech practitioners are currently working on materials and processes that will result in incremental, but nonetheless valuable, improvements to existing technologies including: materials synthesis and fabrication, computation, domestic and commercial lighting, displays, drug-delivery, chemical and biological sensing, bio-mimetic technologies, and medical technologies. Most of these improvements will result from technologies using self assembly and other forms of nanofabrication and synthesis rather than through molecule-by-molecule construction performed by a swarm of nanoassemblers.

[FNa1]. Dr. Martin Moskovits is a Professor of Chemistry and Biochemistry and Dean of the Division of Mathematical and Physical Sciences at the University of California Santa Barbara. He may be reached at mmoskovits @ltsc.ucsb.edu.

[FN1]. See Mick Wilson, et al., Nanotechnology: Basic Science and Emerging Technologies 3-7 (2002); Geoff. A. Ozin & Andre C. Arsenault, Nanochemistry: A Chemistry Approach to Nanomaterials (2005).

[FN2]. K. Eric Drexler, Engines of Creation: The Coming Era of Nanotechnology (1986).

[FN3]. Michael Crichton, Prey (2002).

[FN4]. K. Eric Drexler, Op-Ed., Today's Visions of the Science of Tomorrow, N.Y. Times, Jan. 4, 2003, § A, at 11.

[FN5]. See Vicki L. Colvin, The Potential Environmental Impact of Engineered Nanomaterials,21 Nature Biotech. 1166, 1166-1170 (2003).[FN6]. See Andrew Maynard, Is Nanotechnology Hazardous to Your Health? Machine Design, Dec. 14, 2007, at 71; Davis J. Michael, How to Assess the Risks of Nanotechnology: Learning from Past Experience, 7 J. Nanosci. & Nanotech. 402 (2007).

[FN7]. See A. Welford Castleman, Jr. & Kit Hansell Bowen, Jr.,Clusters: Structure, Eenergetics, and Dynamics of Intermediate States of Matter, 100 J. Phys. Chem. 12911 (1996).

[FN8]. See, e.g., Harold W. Kroto et al., C60 - Buckminsterfullerene, 318 Nature 162 (1985); T. G. Dietz et al., Laser Production of Supersonic Metal Cluster Beams, 74 J. Chem. Phys . 6511 (1981).

[FN9]. See Thomas A. Witten & Philip A. Pincus, Structured Fluids: Polymers, Colloids, Surfactants (2004).

[FN10]. See Marc J. Madou, Fundamentals of Microfabrication: The Science of Miniaturization (2nd ed. 2002).

[FN11]. See Michael Rubinstein & Ralph H. Colby, Polymer Physics (2003).

[FN12]. See George M. Whitesides et al., Molecular Self-Assembly and Nanochemistry: A Chemical Strategy for the Synthesis of Nanostructures,254 Science 1312 (1991); See also Younan Xia & George M. Whitesides, Soft Lithography, 28 Ann. Rev. of Materials Sci. 153 (1998).

[FN13]. Robert G. Parr & Yang Weitao, Density-Functional Theory of Atoms and Molecules (reprint ed. 1994).

[FN14]. Richard P. Feynman, Talk at the Annual Meeting of the American Physical Society at the California Institute of Technology: There's Plenty of Room at the Bottom (Dec. 29, 1959), in 23 Eng'g & Sci., Feb. 1960 at 22-36, available at http://engr.smu.edu/ee/smuphotonics/Nano/FeynmanPlentyofRoom.pdf.

[FN15]. Shinsaku Fujita, Organic Chemistry of Photography 75-82 (2004).

[FN16]. See Ib Chorkendorff et al., Concepts of Modern Catalysis and Kinetics 4-6 (2003).

[FN17]. See Surface-Enhanced Raman Scattering: Physics and Applications 1-13 (Katrin Kneipp et al. eds., 2006).

[FN18]. See Mick Wilson et al., Nanotechnology: Basic Science and Emerging Technologies (2002) 5-8; Geoff. A. Ozin & Andre C. Arsenault, Nanochemistry: A Chemistry Approach to Nanomaterials (2005).

[FN19]. K. Eric Drexler, Op-Ed., Today's Visions of the Science of Tomorrow, N.Y. Times, Jan. 4, 2003, § A, at 11.

[FN20]. K. Eric Drexler & Richard Smalley, Nanotech, 81 Chem. & Eng'g News 37, 37-42 (2003) (open letters addressed to each other).

[FN21]. Id.

[FN22]. See Maxwell's Demon: Entropy, Information, Computing 4 (H. S. Leff & A. F. Rex eds., Institute of Physics Publishing 1990) (reprint of experiment).

[FN23]. Richard P. Feynman et al., The Feynman Lectures on Physics (commemorative ed. 1989) (three volume set).

[FN24]. Id. at 46 (volume I).

[FN25]. See Marc J. Madou, Fundamentals of Microfabrication: The Science of Miniaturization (2nd ed. 2002).

[FN26]. K. Eric Drexler & Richard Smalley, Nanotech, 81 Chem. & Eng'g News 37, 37-42 (2003).

[FN27]. See Vicki L. Colvin, The Potential Environmental Impact of Engineered Nanomaterials, 21 Nature Biotech. 1166, 1166-1170 (2003).

Snow, Jennifer and Giordano, James, "Commentary, Aerosolized Nanobots: Parsing Fact from Fiction for Health Security—A Dialectical View," Health Security,17:1 (2019), Mary Ann Liebert, Inc. DOI: 10.1089/hs.2018.0087. [Excerpt].

Commentary
Aerosolized Nanobots: Parsing Fact from Fiction for Health Security—A Dialectical View
Jennifer Snow and James Giordano

It was recently reported that nanobot sensors could be aerosolized and deployed for the detection of various airborne chemicals.1 Such capabilities are of evident utility in and benefit to medicine, as well as to detect toxins in the environment (functioning as a nanoscalar ''canary'' to warn of hazardous contamination in industrial sites) and/or as a threat awareness system that could be employed in both public and military settings. (2)

Nanoscalar robotics can be used as both sensors and receiver-delivery devices, and the controllability of these technologies enable their directed activity in biological organisms. Such devices—either operating in tandem as distinct sense-and-engage systems, or as single devices with both sense and delivery modes—could be employed to assess, respond to, or modify molecular and chemical characteristics of a biological target. As recent studies have indicated, these approaches can be used in clinical care to more precisely monitor tissue, organ, and overall bodily states and to alter the structure and function of biological tissues and systems at a variety of scales, from the subcel- lular to the systemic and organismic. To be sure, there is significant value in this technology's current and near-term capabilities in affording more granular methods and tools of evaluating and treating disease and injury.(1-3)

However, we posit that the development of aerosolizable nanomaterials and devices also poses defined risks to public health and biosecurity that warrant consideration, address, and constraint. Aerosolized nanobots could be used to sidestep extant proscriptions of the current Biological and Toxin Weapons Convention (BWC) or Chemical Weapons Convention (CWC). (4,5) The properties of these devices that allow their stable aerosolization also confer ability to remain suspended for longer periods of time in a variety of environments. They can be partially or fully autonomous and are capable of storing information with potential to identify or affect specific biological targets. They possess the ability to move independently and up to 2 feet multidirectionally in a closed space, and they can be disseminated much further when dispersed via a spray mechanism or other propellant. Their size (and ''programmability'') allows them to easily enter unpro- tected bodily spaces and to penetrate protective gear. A key limiting factor is the energetics required for nanobots' operations. If the nanobot was to rely on stored energy (eg, that when assembled or released), then energy demand would constrain functional durability, as current nanobotic systems do not have extensive energy-storing capacity.

However, a nanobotic system capable of collecting energy from either its environment (eg, via thermal transfer or conversion), or through interaction with non-robotic nanomaterials, could effectively decrease such constraints. As well, the convergence of nanotechnology with synthetic biology (eg, CRISPR-Cas9 gene editing,

(6,7) use of information about synthesizing viruses (8,9) could lead to a more effective capability to deliver new, and increasingly potent, morbid or lethal synthetic microbes or chem-bio hybrids. These could be customized to create novel agents that could be weaponized and, given their novelty, are not surveilled or recognized by existing regulatory bodies or anticipated by public health and biosecurity operations.

. . .

Of course, it could be argued that although nanotechnology, unlike biological systems, is human-designed and therefore perhaps more programmable, it will, like any other highly distributed information system, nonetheless suffer from unpredictable dynamics. But such unpredictability may confer benefit if and when a range of effects is desired. Moreover, testing the technology in a variety of environmental conditions can decrease both uncertainty and variability of such devices' functional behavior(s). As well, given that nanodevices cannot self-replicate (at least at present), even a modest rate of their destruction could negate their viability. But this may be moot; if the effective potency of the nanodevices to incur a disruptive or destructive effect is sufficiently high, it may be that only ''a little'' is required to do ''a lot,'' and if a great enough number of nanodevices is delivered, then this could account for relative attrition and still leave enough to ''do the job.''

Perhaps nanodevices are not (yet) ready for ''prime time'' use as weaponized agents.(10,11) Yet, it is important to note that the aforementioned constraints can be viewed as challenges to overcome, so that opportunities for creating novel weapons can be exploited. Such possible trajectories should be recognized and regarded. Thus, as have others, (4,5,7) we too reiterate a call for the review, redress, and in some cases revision or reformulation of key guidelines, surveillance, control, restrictions, and enforceable penalties to prevent nefarious development and use of these technological advancements.

In his 2002 science fiction novel, Prey, author Michael Crichton depicted a terrifying view of runaway effects of convergent nanotechnology and genetic engineering.(12) While it is cavalier to look to science fiction scenarios to portend scientific fact, it must be acknowledged that such stories can serve to convey ideas, insights, and cautions, both about science and, perhaps, even more about the ways that individuals and societies view scientific and technological advancement and what such stances reveal. (13) In our view, caution—and preparedness—need not focus on the ease of nanobots going rogue, but on the relative ease and ardor with which nations and rogue actors could go ''nano'' in creating weapons that threaten public safety and biosecurity.

References

1.   Koman VB, Liu P, Kozawa D, et al. Colloidal nanoelectronic state machines based on 2D materials for aerosolizable electronics. Nat Nanotechnol 2018;13(9):819-827.

2.   Ahmad J, Akhter S, Rizwanullah M, et al. Nanotechnology- based inhalation treatments for lung cancer: state of the art. Nanotechnol Sci Appl 2015;8:55-66.

3.   Giordano J, Akhouri R, McBride D. Implantable nano- neurotechnological devices: consideration of ethical, legal, and social issues and implications. J Long Term Eff Med Implants 2009;19(1):83-93.

4.   Wallach E. A tiny problem with huge implications—nanotech agents as enablers or

substitutes for banned chemical weapons: is a new treaty needed? Fordham Int Law J 2009;33(3):858- 956.

5.	Gerstein D, Giordano J. Re-thinking the Biological and Toxin Weapons Convention? Health Secur 2017;15(6):638- 641.

6.	DiEuliis D, Giordano J. Why gene editors like CRISPR/Cas may be a game-changer for neuroweapons. Health Secur 2017;15(3):296-302.

7.	DiEuliis D, Giordano J. Gene editing using CRISPR/Cas9: implications for dual-use and biosecurity. Protein Cell 2018; 9(3):239-240.

8.	Soucheray S. Step-by-step horsepox study stokes dual-use controversy. CIDRAP News January 23, 2018. http://www. cidrap.umn.edu/news-perspective/2018/01/step-step-horse pox-study-stokes-dual-use-controversy. Accessed January 9, 2019.

9.	Snow JJ, Giordano J. Public safety and national security implications of the horsepox study. Health Secur 2018;16(2): 140-142.

10.	Ramsden J. Nanotechnology for military applications. Na- notechnol Percept 2012;8(2):99-131.

11.	Nichols G. Nanotechnology and the new arms race. HDIAC J 2017;4(2):1-6.

12.	Crichton M. Prey. New York: Harper-Collins; 2002.

13.	Wurzman R, Yaden D, Giordano J. Neuroscience fiction as eidola´: social reflection and neuroethical obligations in de- pictions of neuroscience in film. Camb Q Health Care Ethics 2017;26(2):292-312.

XIV.   RESEARCH NOTES on Potentially Applicable Law

A. Federal Law and Criminal Liability, Remedies

Civil Liability
CERCLA (Superfund)
Sec. 107(a) "costs of removal or remedial action" incurred by the federal government, "other necessary costs of response" incurred by any other person, and "damages for injury to, destruction of, or loss of natural resources, including the reasonable costs of assessing ushc injury, destruction, or loss"

SARA (Superfund Amendments)
Health effects and health assessments are recoverable

However, private individuals may not recover compensation for lost property values or individual medical monitoring. *Price v. U.S. Navy*, 39 F.3d 1011 (9th Cir. 1994).

107(f) liability for injury to natural resources is owed to the United States government, to any state for resources "within the State or belonging to, managed by , controlled by, or appertaining to such State," and to any Indian tribe.

122(a) settlement agreements
122(g)(1) *de minimis* settlements, expedited settlements for small-volume waste (4) covenants not to sue, protecting PRPs who settle from future liability to the US related to the Hazardous substance release addressed by a remedial action.

RCRA
3008(d) fines and imprisonment for any person who:
(1) knowingly transports or causes to be transported any hazardous waste identified or listed under this subchapter to a facility which does not have a permit under this subchapter. . .
(2) knowingly treats, stores, or disposes of any hazardous waste identified or listed under this subchapter ---
    (A) without a permit under this subchapter . . .;
    (B)) in knowing violation of any material condition or requirement of such permit . .

    US v. Laughlin, 10 F.3d 961 (2d Cir. 1993)

Jury instructions:
That defendant knowingly disposed of creosote sludge and knew that it "had the potential to be harmful to others or the environment or, in other words, it was not a harmless substance like uncontaminated water."
"public welfare offense" doctrine:
When knowledge is an element of a statute intended to regulate hazardous or dangerous substances, the Supreme Court has determined that the knowledge element is satisfied upon a showing that a defendant was aware that he was performing the proscribed acts; knowledge of regulatory requirements is not necessary.

Clean Water Act
US v. Sinskey, 119 F.3d 712 (8th Cir. 1997)
309(c)(2)(A), punished anyone who "knowingly violates" 301 or a condition or limitation contained in a permit that implements 301.

Jury instruction: in order for the jury to find defendant guilty of acting "knowingly" the proof had to show that he was "aware of the nature of his acts, perform[ed] them intentionally, and [did] not act or fail to act through ignorance, mistake, or accident."
The Court of Appeals held that the government was not required to prove that defendant knew that his acts violated either the CWA or the corporate employer's NPDES effluent discharge permit, "but merely that he was aware of the conduct that resulted in the permit's violation."

[ignorance is not a defense]

Criminal negligence
United States v. Hanousek, 176 F.3d 1116 (9th Cir. 1999) cert denied, 528 U.S> 1102 (2000), defendant appealed his conviction for negligently discharging a harmful quantity of oil into a navigable water of the US in violation of CWA 309(c)(1)(A) and 311(b)(3)
"the failure to use reasonable care."

Responsible Corporate Officers
RCRA 3008(d)(1) knowingly transporting and causing the transportation of hazardous waste to a facility which did not have a permit. Criminal conviction was reversed because "a mere showing of official responsibility . . . is not an adequate substitute for direct or circumstantial proof of knowledge."

Prosecutorial discretion
Factors identified for case selection
"significant environmental harm"
"culpable conduct"

Environmental harm indicated by these factors:
Actual harm
Threat of significant harm
Failure to report environmental releases, and
A trend or common attitude toward noncompliance within the regulated community

Culpable conduct is based on
1. a history of repeated violations,
2. deliberate misconduct,
3. concealment of misconduct of falsification of records
4. tampering with pollution monitoring or control equipment, or
5. conducting pollution related activities without necessary permits or approvals

B. State Law and Tort Law

Trespass
Strict liability
Emotional distress
Enhanced risk of future illness
Medical monitoring
> *Branch v. Western Petroleum, Inc.* 657 P.2d 267 (Utah 1982) all except enhanced risk and medical monitoring

Emotional distress enhanced risk and medical monitoring
> *Ayers v. Township of Jackson*, 525 A.2d 287 (NJ 1987)

Public nuisances
Village of Wilsonville v. SCA Services, Inc., 426 NE2d 824 (Ill. 1981) ordered to remove a "prospective nuisance"

Restatement (Second) Torts sec. 520 (1977)
Liability exists if the activity is "abnormally dangerous"
Weigh the probability and severity of foreseeable harm, whether the activity is unusual or is in an inappropriate location

State, Dept of Environmental Protection v. Ventron Corp., 468 A.2d 150 (NJ 1983)
Court imposed strict liability for harm caused by toxic substances escaping form a landowner's property.

Restatement (Third) of Torts sec., Liabilitly for Physical Harm, Chapter IV, sec. 20, provides that a defendant who carries on abnormally dangerous activity is subject to strict liability for resulting physical harm.
A risk of physical harm can be highly significant for either
1. because the likelihood of harm is unusually high or
2. because the severity of the harm could be enormous

3 years statute of limitations after EXPOSURE
"discovery" rule 40 states
Period of limitation does not begin to run until plantiff discovers her illness.
10 states; adopted sec 309 CERCLA in 1986: any action brought under state law for personal injury or property damages "which are caused or contributed to by exposoure to any hazardous substance . . released into the environment from a facility," the limitation period for the action shall commence on "the date plantiff knew (or reasonably should have known): that the injury or damages were caused or contributed to by the hazardous substance.

C. Federal Torts

Because the government is protected by sovereign immunity, a statutory exception is required to allow individuals to sue the federal government for recovery for injuries that are a particular standard, like gross negligence. The Federal Tort Claims Act allows suits against the federal government for compensation in limited cases.

Federal Tort Claims Act

The Federal Tort Claims Act, 28 U.S. Code § 2674, provides for certain instances when the United States may be liable for injuries to individuals. The United States shall be liable, respecting the provisions of this title relating to tort claims, in the same manner and to the same extent as a private individual under like circumstances, but shall not be liable for interest prior to judgment or for punitive damages.

If, however, in any case wherein death was caused, the law of the place where the act or omission complained of occurred provides, or has been construed to provide, for damages only punitive in nature, the United States shall be liable for actual or compensatory damages, measured by the pecuniary injuries resulting from such death to the persons respectively, for whose benefit the action was brought, in lieu thereof.

With respect to any claim under this chapter, the United States shall be entitled to assert any defense based upon judicial or legislative immunity which otherwise would have been available to the employee of the United States whose act or omission gave rise to the claim, as well as any other defenses to which the United States is entitled.

Defense
The Feres Doctrine, named for *Feres v. United States*, 340 U.S. 135 (1950), holds that the military is not liable for injuries to members of the armed services under the Federal Tort Claims Act.

D.   National Security Law
Chapter 5. Armed Forces, Civil Disturbances, and National Defense
XV. Security Proceedings and Prosecutions
D. Secrecy Agreements, Discovery of Military Secrets
§ 5:595. Privilege against revealing military secrets
Generally, parties in federal civil litigation may obtain discovery regarding any matter, "not privileged," which is relevant to the subject matter involved in the pending action. There is a well-established governmental privilege which absolutely protects military and state secrets because disclosure would jeopardize the national security. The privilege against revealing military secrets belongs to the government and must be asserted by it; it can neither be claimed nor waived by a private party. It is a privilege not to be lightly invoked.

E.   Freedom of Information Act

The Freedom of Information Act (FOIA) provides for open records available to anyone who requests them for any reason. This can be a route for discovery in deciding what causes of action might be supported by the record. However, there is a national security exception to FOIA, which may be relevant in this case.

The Freedom of Information Act, 5 U.S.C.A. § 552, does not apply to matters that are specifically authorized under criteria established by an executive order to be kept secret in the interest of national defense or foreign policy and are in fact properly classified pursuant to such executive order.

What matters are exempt from disclosure under Freedom of Information Act (5 U.S.C.A. § 552(b)(1)) as "specifically authorized under criteria established by an Executive order to be kept secret in the interest of national defense or foreign policy", 29 A.L.R. Fed. 606.

In *Public Educ. Center, Inc. v. Department of Defense*, 905 F. Supp. 19, 169 A.L.R. Fed. 803 (D.D.C. 1995), where suit was brought to obtain copies of videotapes made during a raid involving United States troops in Mogadishu, Somalia, the court held that the refusal of the Department of Defense to release the tapes was proper under the national security exemption.

F.   International Law:

In order to prevent the proliferation of biological weapons, the BWC incorporates a method for addressing alleged violations. In Article VI, a party alleging violations by another party may file a detailed complaint with the United Nations Security Council. If an investigation is commenced, all parties must cooperate pursuant to Articles VI-VII. These enforcement provisions, however, are criticized for their lack of verification measures.

G.   Forms for addressing military information

Complaint—For injunction against withholding, and for order for disclosure, of military aircraft accident board report. Federal Procedural Forms, L. Ed. § 5:481.

Complaint—For injunction to order disclosure of records—For attorneys' fees and costs. Federal Procedural Forms, L. Ed. § 5:482.

Complaint—For injunction to enjoin denial of fee waiver requests—For declaration that plaintiff eligible for, and entitled to, waiver of search and review fees. Federal Procedural Forms, L. Ed. § 5:483.

Order—For in camera inspection of military report to ascertain what portions, if any, may be exempt from mandatory disclosure. Federal Procedural Forms, L. Ed. § 5:484.

Order—For interim award of attorneys' fees and costs. Federal Procedural Forms, L. Ed. § 5:485.

Judgment—Injunction against withholding of military report or records in ordering disclosure thereof. Federal Procedural Forms, L. Ed. § 5:486.
A.L.R. Library, 169 A.L.R. Fed. 495 (Originally published in 2001)

XIV. CONCLUSION

This trial may have felt very futuristic when you began to read it, but as it unfolded it likely became very real and seemingly plausible. Artificial intelligence is potentially going to cause a major shift in our society, and in combination with nanotechnology can be truly an exhilarating advance in the human condition or a devastating disaster for humankind. This trial opens the possibility of a second chance for humankind, in recognizing some of the hazards of unregulated experimentation. By criminally prosecuting the criminals and recovering compensation from the malfeasors we parse the moral, ethical and legal responsibilities that must be attendant with the development of new weapons systems, using artificial intelligence. What could have been a disaster was eliminated by destroying the laboratory and the artificial intelligence inside. (This solution may be reminiscent for those of you who have read or watched Crichton's *Andromeda Strain* where earth was saved by destroying the laboratory where the alien virus was contained.)

www.ingramcontent.com/pod-product-compliance
Lightning Source LLC
Chambersburg PA
CBHW061618210326

41520CB00041B/7490